Lecture Notes in Mathematics

A collection of informal reports and seminars
Edited by A. Dold, Heidelberg and B. Eckmann, Zürich

Series: Mathematisches Institut der Universität Erlangen-Nürnberg
Advisers: H. Bauer und K. Jacobs

289

Bent Fuglede

Københavns Universitet, København/Danmark

Finely Harmonic Functions

Springer-Verlag
Berlin · Heidelberg · New York 1972

AMS Subject Classifications (1970): 3102, 31B05, 31B15, 31B99, 31C05, 31D05

ISBN 3-540-06005-7 Springer-Verlag Berlin · Heidelberg · New York
ISBN 0-387-06005-7 Springer-Verlag New York · Heidelberg · Berlin

Offsetdruck: Julius Beltz, Hemsbach/Bergstr.

Contents

Introduction

0.1. Definitions. The theory of balayage of measures on a harmonic space Ω permits us, in particular, to define the harmonic measure ϱ_x^V relative to a finely [1] open set V and a point $x \in V$ as follows

$$\varrho_x^V = \varepsilon_x^{\complement V} ,$$

that is, the swept-out of the Dirac measure ε_x on the complement of V.

It is therefore natural to introduce corresponding notions of "fine harmonicity" and "fine hyperharmonicity" for numerical functions f defined in a finely open set $U \subset \Omega$ as follows (Definitions 8.1 and 8.3 below):

A function f is called finely hyperharmonic in U if f is finely lower semicontinuous and $> -\infty$ in U, and if the induced fine topology on U has a basis consisting of finely open sets V of fine closure $\tilde{V} \subset U$ such that

$$f(x) \geqslant \int^* f \, d\varepsilon_x^{\complement V} \quad \text{for every} \quad x \in V.$$

A function f is called finely harmonic in U if f is finite and finely continuous in U, and if the finely open sets V with $\tilde{V} \subset U$ such that f is $\varepsilon_x^{\complement V}$-integrable and

---

1) The qualification fine(ly) refers to the fine topology on Ω, that is, the coarsest topology on Ω making all hyperharmonic functions continuous.

$$f(x) = \int f \, d\varepsilon_x^{\complement V} \qquad \text{for every } x \in V$$

form a basis for the fine topology on U.

In view of the regularity of the fine topology the finely harmonic functions have the usual sheaf property relative to that topology, and so do the finely hyperharmonic functions.

Despite the failure of compactness arguments for the fine topology, [2] it turns out that an extensive theory of finely harmonic and finely hyperharmonic functions can be developed in the framework of Brelot's axiomatic theory in the case denoted by A_1; or - essentially equivalently - of a strong harmonic space Ω in the sense of Bauer and satisfying moreover the domination axiom (D) which plays a crucial rôle. This theory of "fine harmonicity", to be developed in the present memoir, is very analogous to the usual theory of harmonic and hyperharmonic functions on a space Ω of the stated kind.

Only in a few instances (regarding the relations between notions involving fine harmonicity and corresponding notions in the usual theory, see notably §10.13 - §10.17) is it necessary to impose further axioms leading to Case A_2 in the theory of Brelot and Mme Hervé, and thus containing the classical case of a Green space in the sense of Brelot and Choquet, and further that of elliptic second order equations with suitably regular coefficients. The whole theory is, however, new even in the classical, say newtonian case, the notion of fine harmonicity being effectively more general than that of usual harmonicity (Theorem 10.11).

---

2) In the classical case all finely compact sets are finite.

A brief, preliminary exposition of the present work was given by the author at the International Congress of Mathematicians in Nice, 1970. [3)]

0.2. Background material. The theory of fine harmonicity draws upon a substantial part of the theory of harmonic spaces. A survey of relevant parts of this latter theory is given in Chapters I and II. In Chapter I the framework is that of a strong harmonic space in the sense of Bauer. (Even more general spaces could be allowed here.) In Chapter II the domination axiom is added. The following results from these two chapters are more or less new and seem to have some independent interest:

The principle of quasi normality (Theorem 3.10): For any finely continuous function f defined on a finely closed set $A \subset \Omega$ there exists a finely continuous function on the whole of Ω which coincides with f on the base of A. [4)] - This result applies to any strong harmonic space. (Recall that the fine topology is completely regular, but generally not normal.)

The remaining results to be listed here refer to a strong harmonic space Ω satisfying the domination axiom (D), or equivalently a Brelot harmonic space (though not necessarily connected) in the case denoted by A_1, that is, with a countable basis, existence of a potential > 0, and satisfying axiom (D).

---

3) A minor correction to this preliminary exposition will be given in §10.15 below.

4) The base, $b(A)$, of a set $A \subset \Omega$ is the set of points of Ω at which A is not thin (effilé).

For any numerical function $f \geq 0$ on Ω we have (Lemma 4.5)

$$R_f(x) = \max\{f(x),\ \hat{R}_f(x)\} \qquad \text{for all } x \in \Omega .$$

As a consequence we have for any increasing sequence of numerical functions $f_n \geq 0$ on Ω (Theorem 4.6)

$$R_{\sup f_n} = \sup_n R_{f_n} \quad ; \qquad \hat{R}_{\sup f_n} = \sup_n \hat{R}_{f_n} .$$

For any potential p on Ω the following propositions are equivalent (with $\bigwedge$ denoting infimum relative to the lattice of all hyperharmonic functions ≥ 0 on Ω with the pointwise order):

i) $\bigwedge_{\lambda>0} \hat{R}_{(p-\lambda)^+} = 0 .$

ii) $\bigwedge_{\lambda>0} \hat{R}_p^{[p>\lambda]} = 0 .$

iii) p is representable as the sum of a sequence of locally bounded potentials, or even of finite and continuous potentials of compact harmonic support. (Theorem 2.2, §2.5, and Lemma 6.4.)

Following the terminology of Brelot, a potential p having these three equivalent properties is said to be <u>semibounded</u>.

For any numerical function $f \geq 0$ which is finely upper semicontinuous quasi everywhere in Ω and majorized by some semibounded potential we have

$$\bigwedge \{ \hat{R}_{f-\varphi} \mid \varphi \in \mathcal{H}_0 ,\ \ \varphi \leq f \} = 0 .$$

If f is even finely continuous quasi everywhere, then

$$\bigwedge \{ \hat{R}_{|f-\varphi|} \mid \varphi \in \mathcal{C}_0^+ \} = 0 .$$

(Theorem 6.5). - Here $\mathcal{H}_0$ denotes the class of all finite valued functions ≥ 0 which are upper semicontinuous and of compact support in the initial topology on Ω. And $\mathcal{C}_0^+$ is the class of all continuous functions from $\mathcal{H}_0$.

There are similar results in which the "capacity" $f \mapsto \hat{R}_f$ is replaced by $f \mapsto \int \hat{R}_f \, dm$ for a suitable measure m on Ω, or by $f \mapsto R_f(x_0)$ for a given point x_0. All these approximation results depend on the Choquet property for the capacities in axiomatic potential theory as established by Brelot.

Finally, let again $f \geq 0$ be finely upper semicontinuous quasi everywhere in Ω and majorized by a semibounded potential. For any sequence of finely closed sets A_i such that

$$f(x) \leq \int f \, d\varepsilon_x^{A_i} \quad \text{for quasi every} \quad x \in \Omega$$

we have

$$f(x) \leq \int f \, d\varepsilon_x^{\cap A_i} \quad \text{for quasi every} \quad x \in \Omega.$$

This result (Theorem 7.3) contains the essence of the fine boundary minimum principle to be mentioned below. It replaces the simple compactness arguments serving to establish the boundary minimum principle for ordinary hyperharmonic functions.

The following special case of Theorem 7.3 is used in the proof thereof:

For any semibounded potential p and any sequence of finely closed sets A_i such that $\hat{R}_p^{A_i} = p$ we have $\hat{R}_p^{\cap A_i} = p$ (Lemma 7.1).

0.3. Basic results on finely harmonic and finely hyperharmonic functions. In describing the similarity between the theory of fine harmonicity to be developed in Chapters III and IV below, and the usual theory of harmonicity, it is illuminating to exhibit clearly the main differences between the two theories.

First, the rôle of the relatively compact, open sets in the usual theory will be taken over now by those finely open sets V (of fine closure $\tilde{V}$ contained in the given finely open set U) for which the finely harmonic (resp. finely hyperharmonic) function f (in U) to be considered is majorized in absolute value (resp. in its negative part) by some finite and continuous potential p on Ω , or more generally by some finite and semibounded potential on Ω . (In some contexts the finiteness of such a semibounded potential may be dropped.)

These sets V form a basis for the fine topology on U (§8.4), but this basis depends on the function f in question (unlike the basis for the initial topology on an open set formed by the open subsets of compact closure contained therein). It turns out (Theorem 9.4 and Lemma 9.5) that this particular basis for the fine topology on U could be used in the above definition of fine harmonicity, resp. fine hyperharmonicity, of the given function f . As a corollary note that a function f is finely harmonic in U if and only if f is both finely hyperharmonic and finely hypoharmonic in U . Moreover, the finely hyperharmonic functions in U form a convex cone, stable under pointwise supremum for increasing nets, and also under pointwise infimum for finite families. The finely harmonic functions in U form a vector space.

A crucial step in developing the present theory was that of establishing the fine boundary minimum principle (Theorem 9.1). A substantial part of Chapters I and II (notably §2, §6, and §7) is devoted to preparing the proof of this theorem. Basically, the method employed is parallel to the sweeping-out process devised by Poincaré for solving the classical Dirichlet problem (cf. particularly Lemma 7.2).

Another basic result (Theorem 9.10) asserts that every finely hyperharmonic function is finely continuous. Thus there is no new "fine-fine" topology. This is tied up with a local extension theorem for finely hyperharmonic functions (Theorem 9.9).

As a consequence, the pointwise limit of a monotone net of finely harmonic functions is finely harmonic if finite (Theorem 9.11). Unlike the situation for usual harmonic functions, such a monotone limit may be infinite precisely at a single point (in the classical case), see Remark 10.7.

A finely hyperharmonic function is said to be finely superharmonic if it is finite quasi everywhere in its domain. Any finite valued, finely superharmonic function $\geqslant 0$ in a finely open set U admits a unique decomposition as the sum of a finely harmonic function in U and a fine potential relative to U, that is, a finely superharmonic function $\geqslant 0$ for which every finely hypoharmonic minorant is $\leqslant 0$ (Theorem 10.7). The finiteness condition cannot be removed.

There is, however, a more general, unique decomposition of any finely superharmonic function $\geqslant 0$ into the sum of a stable fine potential p and a finely continuous numerical function $h \geqslant 0$ which is finely harmonic quasi everywhere in U (hence actually everywhere in $[h < +\infty]$ (Theorem 10.10).

A fine potential p relative to U is called stable if its restriction to $[p < +\infty]$ (and hence to $U \setminus e$ for any polar set e) is again a fine potential (relative to this smaller set). An equivalent property is that of being representable as the sum of a sequence of finite, fine potentials (Theorem 11.18).

In the case of an open set U in the initial topology, those finely hyperharmonic functions in U which are locally bounded from below are precisely the usual hyperharmonic functions in U (Theorem 9.8). The stable fine potentials relative to the open set U are precisely the semibounded (ordinary) potentials relative to U (Theorem 10.12). Any fine potential relative to an open set U is an ordinary potential relative to U. The converse holds, e.g., in the greenian case (Theorem 10.13), but not in a general strong harmonic space satisfying axiom (D).

Via the notion of a nearly finely hyperharmonic function in a finely open set U, the balayage relative to such a set is studied, and the appropriate convergence theorem is proved (Theorem 11.9). The usual rules for swept-out and reduced functions carry over (§11.4 - §11.15). Finally, the specific order on the cone of finely hyperharmonic functions in U is studied, and the specific multiplication is introduced in the case of stable fine potentials (§11.16 - §11.22).

0.4. Main examples. For any numerical function f defined in a base [5] B, the function u defined in all of Ω by

$$u(x) = \int^* f \, d\varepsilon_x^B , \qquad x \in \Omega ,$$

is finely harmonic in the finely open set of all points $x \in \complement B$ such that $u(x)$ is finite (Theorem 9.13).

For any function $f \geq 0$ on Ω, $\hat{R}_f$ is finely harmonic in every finely open subset of $[\hat{R}_f < +\infty]$ in which f is finely hypoharmonic (for instance equal to 0). (Lemma 9.7.)

In a Green space Ω with Green kernel G the potential $G\mu$ of a positive measure μ is finely harmonic precisely in all finely open sets U such that $\mu^*(U) = 0$ and $G\mu < +\infty$ in U (Theorem 8.10).

0.5. Results involving fine connectivity. The present theory provides a natural means of studying the connectivity properties of the fine topology on our strong harmonic space Ω satisfying axiom (D). In this way we obtain simple proofs that the fine topology on Ω is locally connected (§9.11), and that every subdomain of Ω in the initial topology is likewise a fine domain, that is, a finely connected, finely open set. These results were established by the author in a more direct way in an earlier paper.

Using a result concerning removable singularities for finely harmonic functions (Theorem 9.15), it is easily shown

---

5) A base is a finely closed set which is not thin at any one of its points. The complement of a base is called a regular finely open set.

that $U \setminus e$ is a fine domain for every fine domain U and every polar set e (Theorem 12.2).

As a consequence we obtain, for any set $A \subset \Omega$,

$$i(\partial_f A) = i(A) \cup i(\complement A),$$

$$b(\partial_f A) = b(A) \cap b(\complement A),$$

where $i(A) = A \setminus b(A)$ is the polar set of all points of A at which A is thin, and $\partial_f A$ denotes the fine boundary of A. Hence

$$i(\partial_f A) = i(A) \qquad \text{if } A \text{ is finely closed.}$$

In particular, a finely closed set A is a base (that is, $i(A) = \emptyset$, or equivalently $b(A) = A$) if and only if $\partial_f A$ is a base (Lemma 12.4).

For any <u>finely open</u> set U the points of the polar set $i(\partial_f U) = i(\complement U)$ are called the <u>irregular points</u> of $\partial_f U$, or the irregular (fine boundary) points for U. The set

$$V = U \cup i(\complement U) = \complement b(\complement U),$$

derived from U by adding all the irregular points for U, is the smallest regular finely open set containing U (§12.5). This observation allows us to reduce several problems involving a finely open set U (in particular the fine Dirichlet problem) to the case where U is regular. This represents a simplification as compared to the usual theory because the above extension V is generally not open in the initial topology even though U is supposed to be open.

A finely hyperharmonic function $\geqslant 0$ in a fine domain is either > 0 throughout, or else identically $= 0$ (Theorem 12.6).

A finely hyperharmonic function in a fine domain is either finite quasi everywhere or else identically $+\infty$ (Theorem 12.9). - This latter result depends on the main result, to be described now, concerning the fine support of swept-out measures.

0.6. Applications to the balayage of measures or functions. For any admissible [6] measure μ on Ω and for any base $B \subset \Omega$ such that $\mu(B) = 0$, the swept-out measure μ^B has a smallest finely closed support, viz. the fine boundary of the union of those fine components of $\complement B$ which are non negligible for μ. And the complement of this union is the largest among all bases E such that $\mu^E = \mu^B$. (Theorem 12.7.)

As a consequence it is found, for any fine domain $U \subset \Omega$, that the fine support of the harmonic measure $\varepsilon_x^{\complement U} = \varepsilon_x^{\partial_f U}$ equals the regular part $b(\partial_f U) = \partial_f U \setminus i(\partial_f U)$ of the fine boundary of U, not only if $x \in U$, but also when x is an irregular point for U. (In either case, x is in fact a point of the extension V of U to a regular finely open set described above.) - This result contains and refines the analogous result obtained by Effros and Kazdan for the case of a domain U in the initial topology.

For the balayage of semibounded potentials results are obtained (Theorems 13.3 and 13.4) parallel to those concerning the fine support of swept-out measures. There is also an actual

---

6) A positive measure μ on Ω is called admissible if every finite and continuous potential on Ω of compact harmonic support is μ-integrable (e.g. if μ has compact support).

interrelation between these two types of balayage (Theorem 13.5). As an application of this it is found (Theorem 13.6) that

$$(\mu^B)^A = \mu^A \iff \mu^{A\cup B} = \mu^B$$

for any admissible measure μ and any two sets $A, B \subset \Omega$. This bi-implication, in the case $\mu = \varepsilon_x$, was suggested by Nguyen-Xuan-Loc, who also gave an independent, probabilistic proof for that case.

0.7. The fine Dirichlet problem. It is shown how the (proper) fine Dirichlet problem, suitably defined, admits a unique solution for any regular finely open set U. It is understood that the prescribed finely continuous function f on $\partial_f U$ as well as the solution are required to be majorized in absolute value by some finite and semibounded potential on Ω. This solution u is given by

$$u(x) = \int f \, d\varepsilon_x^{\complement U} = \int f \, d\varepsilon_x^{\partial_f U}$$

(Theorem 14.1). The case of an irregular finely open set U is reduced to the regular case in the usual way, passing to the regular finely open set $V = U \cup i(\complement U)$, and using a result (Theorem 9.14) concerning removable singularities for finely hyperharmonic functions. - For an irregular fine domain U this leads to an extension and refinement of a result for a usual irregular domain obtained by Effros and Kazdan concerning the antilattice property of the set of boundary functions for which the Dirichlet problem is properly solvable.

The method of Perron-Wiener-Brelot carries over easily to the generalized fine Dirichlet problem (Theorem 14.6). When applied to an open set U in the initial topology, this leads to an extension of a result of Brelot according to which fine limits may be used equally well in the definition of the upper and lower generalized solution of the Dirichlet problem.

0.8. Application to the study of the Choquet property. For any numerical function $f \geq 0$ on Ω it is shown (Theorem 15.1) that if the set function $E \mapsto \hat{R}_f^E$ has the Choquet property

$$\bigwedge \{ \hat{R}_f^{A \setminus F} \mid F \text{ closed, } F \subset A \} = 0$$

for every finely closed set $A \subset \Omega$, then so has each of the set functions $E \longmapsto \int \hat{R}_f^E \, dm$ (m any positive measure on Ω) and $E \longmapsto R_f^E(x_0)$ $(x_0 \in \Omega)$, if it is finite on compact sets.

According to Brelot's study of the Choquet property in axiomatic potential theory, this result applies to any locally bounded f, and in the classical case even to any f majorized locally by some superharmonic function which is integrable with respect to m, resp. finite at x_0. In each of these situations this leads to the Choquet property for the capacity $\int \hat{R}_f^{\cdot} \, dm$, resp. $R_f^{\cdot}(x_0)$, even if m is allowed to charge polar sets, and x_0 to be polar.

0.9. Suggested plan of reading. The reader with some background in axiomatic potential theory may pass directly to sections 7 - 9, consulting the preceding sections on the way. Attention is called, however, to Definition 4.10. The remaining sections 10 - 15 are largely independent of one another.

0.10. Acknowledgements. In §10 and §11 I have benefited much from having had access to the manuscript to the forthcoming book: Potential Theory on Harmonic Spaces, by C. Constantinescu and A. Cornea. I am indebted to C. Berg for reading the original manuscript of the present work and pointing out various inaccuracies, and to Nguyen-Xuan-Loc for suggesting certain improvements.

CHAPTER I

Preliminaries

In this chapter Ω denotes a strong harmonic space in the sense of Bauer [1].

1. The cone of positive hyperharmonic functions

We denote throughout by $\mathcal{U}$ the convex cone of all hyperharmonic functions $\geqslant 0$ on the strong harmonic space Ω , and by $\mathcal{S}$ the face (= hereditary convex subcone) of $\mathcal{U}$ formed by the superharmonic functions $\geqslant 0$ on Ω . - For the results of the present section we generally refer to [1], [3], or [22 bis].

1.1. It is well known that $\mathcal{U}$ is a complete lattice with respect to the natural pointwise order $\leqslant$ on $\mathcal{U}$ as inherited from the complete lattice

$$\mathcal{F}^{+} = [0, +\infty]^{\Omega}$$

of all numerical functions $\geqslant 0$ on Ω .

Following [3] we denote by $\bigvee$ and $\bigwedge$ the supremum and the infimum, respectively, for subfamilies of $\mathcal{U}$ and taken relatively to $\mathcal{U}$ with the pointwise order. The corresponding notions relative to all of $\mathcal{F}^{+}$ are denoted by sup and inf , respectively. Recall that

$$\bigwedge_i u_i = \widehat{\inf_i u_i}$$

for any family (u_i) on $\mathcal{U}$. As usual, $\hat{f}$ denotes, for any function $f \in \mathcal{F}^{+}$, the l.s.c. envelope of f (the greatest lower semicontinuous minorant for f). By Choquet's topological lemma there is a countable subfamily (u_{i_n}) such that $\bigwedge_n u_{i_n} = \bigwedge_i u_i$.

For any pointwise upper directed family (u_i) on $\mathcal{U}$ the pointwise supremum $\dot{\sup_i} u_i$ relative to $\mathcal{F}^+$ is again hyperharmonic, and hence is likewise the pointwise supremum of (u_i) relative to $\mathcal{U}$:

$$\bigvee_i u_i = \sup_i u_i \qquad \text{if } (u_i) \text{ is upper directed.}$$

1.2. In addition to the pointwise ordering, $\leqslant$, of $\mathcal{U}$ we have on $\mathcal{U}$ the important specific order, $\preccurlyeq$, defined by

$$[\, u_1 \preccurlyeq u_2 \,] \Longleftrightarrow [\, \exists\, u \in \mathcal{U} : \quad u_1 + u = u_2 \,]$$

for $u_1, u_2 \in \mathcal{U}$. Clearly $u_1 \preccurlyeq u_2$ implies $u_1 \leqslant u_2$. If u_1 is superharmonic ($u_1 \in \mathcal{S}$), then u is uniquely determined here by u_1 and $u_2 = u_1 + u$, and we may hence define $u_2 - u_1 = u$. This follows, e.g., from [1, p. 152] or [3, Lemma 2.1].

1.3. $\mathcal{U}$ is a complete lattice likewise with respect to the specific order. The supremum and infimum with respect to the specific order are denoted by $\curlyvee$ and $\curlywedge$, respectively, as in [3].

For any upper directed family (u_i) on $\mathcal{U}$ with the specific order,

$$\curlyvee u_i = \bigvee u_i = \sup u_i .$$

For any specifically lower directed family (u_i) on $\mathcal{U}$,

$$\curlywedge u_i = \bigwedge u_i = \widehat{\inf u_i} .$$

For these results see [3, Lemma 1.9], [22 bis, Chap. IV].

1.4. The cone $\mathcal{S}$ of superharmonic functions ≥ 0 on Ω is a <u>face</u> in $\mathcal{U}$ with the pointwise order, a fortiori with respect to the specific order. This means that any $u \in \mathcal{U}$ having a majorant of class $\mathcal{S}$ is itself of class $\mathcal{S}$. It follows that $\mathcal{S}$ is a conditionally complete lattice in either ordering.

1.5. Using the uniqueness property in §1.2 one easily obtains, for any family (u_i) on $\mathcal{U}$ and any $s \in \mathcal{S}$,

$$\Upsilon_i (s + u_i) = s + \Upsilon_i u_i \; ; \quad \bigwedge_i (s + u_i) = s + \bigwedge_i u_i . \tag{1}$$

For any family (u_i) on $\mathcal{S}$ with a specific majorant $u \in \mathcal{S}$ it follows that

$$u = \Upsilon_i u_i + \bigwedge_i (u - u_i) . \tag{2}$$

In particular, for any $s, t \in \mathcal{S}$,

$$s + t = s \curlyvee t + s \curlywedge t . \tag{3}$$

1.6. The lattice cone $\mathcal{S}$ with the specific order has the <u>Riesz decomposition property</u>:

For any $s, s_1, s_2 \in \mathcal{S}$ such that $s \preccurlyeq s_1 + s_2$ there exist $u_1, u_2 \in \mathcal{S}$ such that $u_1 \preccurlyeq s_1$, $u_2 \preccurlyeq s_2$, and $s = u_1 + u_2$. For instance, one may take $u_1 = s \curlywedge s_1$ and hence $u_2 = s - s \curlywedge s_1$. In fact, $u_2 \preccurlyeq s_2$ because

$$(s \curlywedge s_1) + u_2 = s \preccurlyeq (s + s_2) \curlywedge (s_1 + s_2) = (s \curlywedge s_1) + s_2$$

on account of (1) above. (The possibility of subtracting $s \curlywedge s_1$ on both sides again stems from the uniqueness result in §1.2.)

Clearly this decomposition property extends to finite sums of more than two functions $s_i \in \mathcal{S}$.

Essentially equivalent to the preceding results is the well known fact that $\mathcal{S}$, with the specific order, can be imbedded as the positive cone in a conditionally complete vector lattice.

1.7. We shall consider the following subcones of $\mathcal{S}$:

$\mathcal{P}$ = the cone of all potentials (on Ω),

$\mathcal{P}^s$ = the semibounded potentials (cf. §2),

$\mathcal{P}^b$ = the locally bounded potentials,

$\mathcal{P}^c$ = the finite and continuous potentials,

$\mathcal{P}_0^c$ = the potentials of class $\mathcal{P}^c$ harmonic off some compact set.

Note that

$$\mathcal{P}_0^c \subset \mathcal{P}^c \subset \mathcal{P}^b \subset \mathcal{P}^s \subset \mathcal{P} \subset \mathcal{S} \subset \mathcal{U} .$$

Each of these convex cones is a face in $\mathcal{U}$ with respect to the specific order. Except for $\mathcal{P}_0^c$ and $\mathcal{P}^c$ the same holds for the pointwise order.

1.8. For any set $A \subset \Omega$ and any function $f \geq 0$ defined at least in A the <u>reduced</u> and the <u>swept-out</u> function are defined as functions on Ω by

$$R_f^A = \inf \{ u \in \mathcal{U} \mid u \geq f \text{ in } A \} ,$$

$$\hat{R}_f^A = \bigwedge \{ u \in \mathcal{U} \mid u \geq f \text{ in } A \} ,$$

respectively. Thus $\hat{R}_f^A$ is the l.s.c. envelope of R_f^A . The abbreviations $R_f^\Omega = R_f$ and $\hat{R}_f^\Omega = \hat{R}_f$ are used.

1.9. A superharmonic function $p \in \mathcal{S}$ is a <u>potential</u> if and only if

$$\inf \{ R_p^{\complement K} \mid K \text{ compact} \} = 0 .$$

For any compact set K , $\hat{R}_1^K$ is a locally bounded potential: $\hat{R}_1^K \in \mathcal{P}^b$. For any finite and continuous function $f \geq 0$ on Ω of compact support it is shown in [4, Lemma 4.2] that

$$R_f = \hat{R}_f \in \mathcal{P}_0^c .$$

1.10. The <u>harmonic support</u>, $S(p)$, of a numerical function p on Ω is defined as the smallest closed set such that p is harmonic in the complement.

<u>Definition</u>. A superharmonic function $p \geq 0$ on Ω is said to have the <u>domination property</u> if the implication

$$[u \geq p \text{ on } S(p)] \Longrightarrow [u \geq p \text{ in } \Omega]$$

holds for every $u \in \mathcal{U}$.

Equivalently: $R_p^S = p$, or equally well $\hat{R}_p^S = p$, where $S = S(p)$. (In fact, $\hat{R}_p^S \leq R_p^S \leq p$ for any $p \in \mathcal{S}$, and p is l.s.c.)

Every finite and continuous potential p has the domination property. This follows from the boundary minimum principle in its general form, see e.g. [1, Kor. 2.4.3], applied to $u - p$ which is hyperharmonic in $\complement S(p)$, l.s.c. and $\geq -p$ everywhere, and ≥ 0 on the boundary of $\complement S(p)$.

2. Semibounded potentials

2.1. Definition (Brelot [16, p. 41]). A potential p on Ω is said to be semibounded if

$$\bigwedge_{\lambda} \hat{R}_p^{[p>\lambda]} = 0$$

as λ ranges over the real constants. (Here $[p>\lambda]$ denotes the set of points $x \in \Omega$ such that $p(x) > \lambda$.)

We denote by $\mathcal{P}^s$ the class of all semibounded potentials on Ω . Clearly $\mathcal{P}^s$ is hereditary in $\mathcal{U}$, that is, from $u \in \mathcal{U}$, $p \in \mathcal{P}^s$, and $u \leqslant p$ follows $u \in \mathcal{P}^s$.

Every locally bounded potential p is semibounded, that is, $\mathcal{P}^b \subset \mathcal{P}^s$. For any compact set K we have, in fact, $[p>\lambda] \subset \complement K$ for all sufficiently large λ , and so by §1.7

$$\inf_{\lambda} R_p^{[p>\lambda]} \leqslant \inf_{K} R_p^{\complement K} = 0 . \tag{4}$$

2.2. Theorem. The following 3 properties are equivalent for a superharmonic function $p \geqslant 0$ on Ω :

i) p is a semibounded potential: $p \in \mathcal{P}^s$.

ii) p is the specific (or equivalently, the pointwise) supremum of its specific minorants of class $\mathcal{P}^b$:

$$p = \Upsilon \{ q \in \mathcal{P}^b \mid q \preccurlyeq p \},$$

or equivalently

$$\curlywedge \{ p - q \mid q \in \mathcal{P}^b,\ q \preccurlyeq p \} = 0 .$$

iii) p is representable as the pointwise sum of a sequence of locally bounded potentials:

$$p = \sum p_n , \qquad p_n \in \mathcal{P}^b .$$

Proof. Since $\mathcal{P}^b$ is a sublattice of $\mathcal{S}$ in the specific order (being a face in $\mathcal{S}$), $\{q \in \mathcal{P}^b \mid q \preccurlyeq p\}$ is specifically upper directed, and hence $\{p - q \mid q \in \mathcal{P}^b, \ q \preccurlyeq p\}$ is specifically lower directed. Consequently, §1.3 is applicable.

Ad i) $\Longrightarrow$ iii). Let (K_n) denote an increasing sequence of compact sets covering Ω . For any natural number n write

$$E_n = [p > n] \cup \complement K_n .$$

According to Constantinescu [21, Theorem 1.1]

$$p = \hat{R}_p^{\Omega} \preccurlyeq \hat{R}_p^{\complement E_n} + \hat{R}_p^{E_n} ,$$

and consequently, by the Riesz decomposition property (§1.6),

$$\begin{aligned} p &= q_n + r_n , \\ q_n &\leqslant \hat{R}_p^{\complement E_n} \leqslant n \hat{R}_1^{K_n} , \\ r_n &\leqslant \hat{R}_p^{E_n} \leqslant \hat{R}_p^{[p>n]} + \hat{R}_p^{\complement K_n} . \end{aligned}$$

It follows that $q_n \in \mathcal{P}^b$ because $\hat{R}_1^{K_n} \in \mathcal{P}^b$. Moreover,

$$\curlywedge r_n \leqslant \wedge r_n \leqslant \ \inf r_n \leqslant \inf \hat{R}_p^{[p>n]}$$

by (4) because the sequences $(\hat{R}_p^{[p>n]})$ and $(\hat{R}_p^{\complement K_n})$ are pointwise decreasing. Since p is semibounded, we obtain

$$\curlywedge r_n \leq \curlywedge \hat{R}_p^{[p>n]} = 0 ,$$

and so $p = \curlyvee q_n$ by (2), §1.5. Writing $q_n' = q_1 \curlyvee \ldots \curlyvee q_n$, we have

$$p = \curlyvee q_n' = \sup q_n' .$$

Consequently, iii) holds with $p_1 = q_1'$, $p_n = q_n' - q_{n-1}'$ for $n \geq 2$ because $q_n' \in \mathcal{P}^b$ and $q_n' \succcurlyeq q_{n-1}'$, whence $p_n \in \mathcal{P}^b$.

Ad iii) $\Longrightarrow$ ii). Writing $q_n = p_1 + \ldots + p_n$ $(\in \mathcal{P}^b)$, we infer from iii)

$$p = \sup q_n = \curlyvee q_n$$

in view of §1.3 because (q_n) is specifically increasing.

Ad ii) $\Longrightarrow$ i). Let $q \in \mathcal{P}^b$ and $q \preccurlyeq p$. Then $r :=$ $p - q \in \mathcal{S}$. For any $\lambda > 0$ we have

$$[p > 2\lambda] \subset [q > \lambda] \cup [r > \lambda].$$

For any $u \in \mathcal{U}$ such that $u \geq q$ on $[q > \lambda]$ we obtain

$$p \leq u + 2r \qquad \text{on} \qquad [p > 2\lambda].$$

At any point of $[q > \lambda]$ we have, in fact, $u \geq q$ and hence $p = q + r \leq u + r$. And at any point of $[p > 2\lambda] \cap [q \leq \lambda]$ we have $p > 2q$ and hence $p < 2p - 2q = 2r$.

Altogether we have established that $R_p^{[p>2\lambda]} \leq u + 2r$ for any $u \in \mathcal{U}$ such that $u \geq q$ on $[q > \lambda]$. Taking pointwise infimum over these u, we further obtain

$$R_p^{[p>2\lambda]} \leq R_q^{[q>\lambda]} + 2r ,$$

and consequently, in view of [1, p. 48] or [3, Lemma 1.3],

$$\hat{R}_p^{[p>2\lambda]} \leq \hat{R}_q^{[q>\lambda]} + 2r. \tag{5}$$

It follows that

$$\bigwedge_\lambda \hat{R}_p^{[p>2\lambda]} \leq 2r$$

because $q \in \mathcal{P}^b \subset \mathcal{P}^s$. Making q range over $\{q \in \mathcal{P}^b \mid q \prec p\}$, we obtain from ii) that $\bigwedge r = \bigwedge(p-q) = 0$, and we conclude that $\bigwedge \hat{R}_p^{[p>2\lambda]} = 0$. Finally, p is a potential by [22 bis, Prop. 2.2.2] according to which $\mathcal{P}$ is a band in $\mathcal{S}$ with the specific order. ▯

2.3. The last characterization implies that $\mathcal{P}^s$ is a convex cone (subcone of $\mathcal{P}$ and hence of $\mathcal{U}$). Being also hereditary in $\mathcal{U}$, $\mathcal{P}^s$ is therefore a <u>face</u> of $\mathcal{U}$ in the pointwise as well as the specific ordering. In particular, $\mathcal{P}^s$ is a conditionally complete lattice in either ordering, just like $\mathcal{S}$, $\mathcal{P}$, and $\mathcal{P}^b$. Like $\mathcal{P}$, $\mathcal{P}^s$ is even a <u>band</u> in $\mathcal{S}$ with the specific order. It follows, in fact, from §2.4 below that every subset of $\mathcal{P}^s$ which is specifically majorized in $\mathcal{S}$, has its specific supremum relative to $\mathcal{S}$ in $\mathcal{P}^s$.

2.4. The above theorem remains valid if $\mathcal{P}^b$ is replaced throughout by the larger cone $\mathcal{P}^s$. (The proof of iii) $\Longrightarrow$ ii) $\Longrightarrow$ i) carries over, while i) $\Longrightarrow$ iii) becomes trivial.)

2.5. As a comment in the opposite direction note that, under the additional hypothesis of the <u>domination</u> <u>axiom</u> (see Chapter II), Theorem 2.2 would remain valid if $\mathcal{P}^b$ were replaced by the <u>smaller</u> cone $\mathcal{P}_0^c$ of all finite and continuous potentials of compact harmonic support.

To see this, note first that, by [5, Théorème 1], $\mathcal{P}^b$ is contained in the band $\mathcal{M}$ (in $\mathcal{S}$) formed by all $s \in \mathcal{S}$ which are representable as the sum of a sequence of finite continuous functions of class $\mathcal{S}$. According to Theorem 2.2 above this means that $\mathcal{P}^s \subset \mathcal{M}$, and indeed that

$$\mathcal{P}^s = \mathcal{M} \cap \mathcal{P}.$$

Next, in the definition of $\mathcal{M}$ as quoted above, $\mathcal{P}^c$ may be replaced by $\mathcal{P}_0^c$ on account of the partition theorem of Mme Hervé [29, Théorème 12.2], according to which every potential p is the specific supremum of its specific minorants of compact harmonic support, and even of its specific restrictions p_ω to all open, relatively compact sets $\omega \subset \Omega$. In fact, the potential p'_ω defined by $p = p_\omega + p'_\omega$ is harmonic in ω. Hence $\bigwedge_\omega p'_\omega$ is harmonic and $\leq p$ in Ω, and consequently $= 0$.

2.6. Now suppose that Ω is a <u>Green space</u> with the Green kernel G. The potentials on Ω are then represented as $p = G\mu$ where μ ranges over all admissible measures, cf. §3.1. The following known result is mentioned, e.g., in Brelot [16, §5]:

<u>Theorem</u>. Each of the following conditions is necessary and sufficient for a greenian potential $p = G\mu$ to be semi-bounded:

i) μ is representable as the sum of a sequence of measures μ_n (≥ 0) such that $G\mu_n \in \mathcal{P}_0^c$.

ii) μ does not charge the polar sets.

iii) $\mu([G\mu = +\infty]) = 0$.

Proof. It follows from §2.5 above that i) is equivalent to the semiboundedness of $p = G\mu$. The implication i) $\Longrightarrow$ ii) reduces to the case $p = G\mu \in \mathcal{P}_0^c$. For any polar set e there exists a potential $G\lambda$ which equals $+\infty$ in e. It follows that $\mu^*(e) = 0$ because λ is admissible and hence

$$\int G\lambda \, d\mu = \int G\mu \, d\lambda < +\infty .$$

From ii) follows iii) since $[G\mu = +\infty]$ is polar. Finally, iii) implies i) according to a general observation concerning potentials with respect to a kernel satisfying the continuity principle, see Choquet [19, Cor. du lemme 1]. ∎

2.7. It is known that Theorem 2.6 holds more generally for any harmonic space in case A_2 of Brelot's axiomatic theory. This case is obtained from case A_1 (= connected strong harmonic space with axiom (D)) by adding the following 3 further axioms (see Mme Hervé [29, §11]):

a) All potentials on Ω with the same one-point harmonic support are proportional.

b) The topology on Ω has a basis formed by completely determining sets.

c) Thinness and adjoint thinness of a set at a point are identical properties.

As a corollary of the above theorem note that, in the greenian case (or in case A_2), every finite potential is semibounded. This would not hold in general in case A_1, see §10.15 below.

3. Balayage of measures, Base, thinness, and fine topology

3.1. Definition. A measure μ (always positive Radon) on Ω is called admissible if $\int p\,d\mu < +\infty$ for every $p \in \mathcal{P}_0^c$.

Clearly, any measure of compact support is admissible. In the greenian case a measure μ is admissible if and only if its Green potential $G\mu$ is not identically $+\infty$, that is, $G\mu$ should be a potential in the present sense of the axiomatic theory In particular, in newtonian potential theory in $\mathbb{R}^n$, $n \geq 3$, μ is admissible if and only if $\int_{|x| \geq 1} |x|^{2-n} d\mu < +\infty$.

3.2. We refer to Constantinescu [21] for the balayage of admissible measures on the strong harmonic space Ω :

For every admissible measure μ and any set $A \subset \Omega$ there exists precisely one measure μ^A on Ω such that

$$\int u\,d\mu^A = \int \hat{R}_u^A\,d\mu \tag{6}$$

for every $u \in \mathcal{P}_0^c$, and then likewise for every $u \in \mathcal{U}$.

This measure μ^A is called the swept-out, or balayée, of μ on A ; it is likewise admissible (because $\hat{R}_u^A \leq u$).

For any 3 sets A, E, and B such that $A \subset E \subset B$ the relation $\mu^A = \mu^B$ clearly implies $\mu^A = \mu^E$.

In particular, the swept-out ε_x^A of the Dirac measure ε_x at a point $x \in \Omega$ is characterized by

$$\int u\,d\varepsilon_x^A = \hat{R}_u^A(x) \qquad \text{for every } u \in \mathcal{U}. \tag{7}$$

3.3. The <u>base</u> $b(A)$ of a set $A \subset \Omega$ is defined by

$$b(A) = \{ x \in \Omega \mid \varepsilon_x^A = \varepsilon_x \}.$$

A is called <u>thin</u>, or <u>effilé</u>, at a point $x \in \Omega$ if $x \in \complement b(A)$, that is, if $\varepsilon_x^A \neq \varepsilon_x$.[7] For any $u \in \mathcal{U}$, $\hat{R}_u^A = u$ on $b(A)$.

A set $E \subset \Omega$ is called <u>semipolar</u> if E is representable as the union of a sequence of sets E_n such that $b(E_n) = \emptyset$ (that is, E_n should be everywhere thin).

If A_1 and A_2 are thin at x then so is $A_1 \cup A_2$, see [21, Cor. 2.11]. Thus the base operation is <u>additive</u>:

$$b(A_1 \cup A_2) = b(A_1) \cup b(A_2)$$

for any $A_1, A_2 \subset \Omega$. Clearly $A_1 \subset A_2$ implies $b(A_1) \subset b(A_2)$.

3.4. In [21, p. 278] Constantinescu has defined a notion of <u>strict potential</u> and established the existence of a strict potential of class $\mathcal{P}^c$ (on any strong harmonic space). For any prescribed admissible measure μ there even exists a strict potential $p \in \mathcal{P}^c$ such that $\int p\, d\mu < +\infty$, see Constantinescu and Cornea [22 bis, Prop. 7.2.1].[8]

---

7) This definition of thinness was shown by Constantinescu [21] to be equivalent to the corresponding local property ("weak thinness"). It is therefore an extension of the original notion of thinness of a set A at a point $x \in \complement A$ as defined by Brelot in [8] (and earlier papers).

8) Alternatively, let q denote any strict potential of class $\mathcal{P}^c$. Cover Ω by an increasing sequence of compact sets K_n, and choose constants $c_n > 0$ so that $\sum c_n < +\infty$ and $\sum c_n \int \hat{R}_q^{K_n} d\mu < +\infty$. Then $p := \sum c_n \hat{R}_q^{K_n}$ is a strict potential of class $\mathcal{P}^c$ with $\int p\, d\mu < +\infty$.

For any strict potential p the inequality $\hat{R}_p^A(x) < p(x)$ is equivalent to the thinness of the set A at the point x. Thus

$$b(A) = \{x \in \Omega \mid \hat{R}_p^A(x) = p(x)\}. \qquad (8)$$

As a consequence of this, note that $b(A)$ is a G_δ-set. Moreover, $b(\Omega) = \Omega$, that is, the whole space Ω is nowhere thin.

3.5. The <u>fine topology</u> on Ω is defined as the coarsest topology on Ω such that all hyperharmonic functions $u \geq 0$ become continuous. Topological notions with respect to the fine topology are distinguished by the term "fine(ly)" from those pertaining to the initially given topology on Ω.

The fine topology is finer than the initial topology. Every hyperharmonic function is finely continuous.

A set A is <u>thin</u> at a point x not in A if and only if $\complement A$ is a fine neighbourhood of x, that is, if x does not belong to the fine closure of A. In particular, thinness is a <u>local property</u> also in the fine topology.

A set A is thin at a point x belonging to A if and only if $A \setminus \{x\}$ and $\{x\}$ are both thin at x, or equivalently if and only if x is a <u>finely isolated</u> point of A such that $\{x\}$ is thin at x.

3.6. The <u>fine closure</u> of a set $A \subset \Omega$ is denoted by $\tilde{A}$, and the <u>fine boundary</u> of A by $\partial_f A$. Furthermore, we denote by $i(A)$ the set of all finely isolated points x of A such that $\{x\}$ is thin at x. Thus we have

$$\tilde{A} = A \cup b(A) \qquad (\supset b(A)),$$

$$\tilde{A} \setminus A = b(A) \setminus A \; ; \qquad i(A) = A \setminus b(A) ,$$

$$\tilde{A} = b(A) \cup i(A) \; ; \qquad b(A) \cap i(A) = \emptyset .$$

It follows that the finely closed sets are precisely those sets A for which $b(A) \subset A$.

3.7. For any numerical function $f \geq 0$ on Ω with the <u>finely u.s.c. envelope</u> $\tilde{f}$,

$$R_{\tilde{f}} = R_f \; ; \qquad \hat{R}_{\tilde{f}} = \hat{R}_f ,$$

because $u \geq f$ is equivalent to $u \geq \tilde{f}$ for any $u \in \mathcal{U}$ (by the fine continuity of u).

For any <u>finely l.s.c.</u> function $f \geq 0$ on Ω we have

$$\hat{R}_f = R_f$$

In fact, $R_f \geq f$ implies $\hat{R}_f \geq f$ because $\hat{R}_f$ is the l.s.c. envelope of R_f likewise in the fine topology, see e.g. [1] or [4].

Again with f finely l.s.c. we have for any set A

$$R_f^{\tilde{A}} = R_f^A \; ; \qquad \hat{R}_f^{\tilde{A}} = \hat{R}_f^A$$

because $[u \geq f]$ is finely closed for every $u \in \mathcal{U}$.

It follows that the balayage of measures, the notion of thinness, and the base operation are stable under fine closure:

$$\mu^{\tilde{A}} = \mu^A \; ; \qquad b(\tilde{A}) = b(A) \; ; \qquad i(\tilde{A}) = i(A)$$

for any admissible measure μ and any set A. Consequently, $b(b(A)) \subset b(A)$, that is, $b(A)$ is finely closed.

Since $i(\Omega) = \emptyset$ by §3.4, we get for any <u>finely</u> <u>open</u> set U

$$i(U) = \emptyset ; \qquad b(U) = \tilde{U} \quad (\supset U).$$

For arbitrary sets A we therefore obtain

$$i(A) = i(\tilde{A}) \subset i(\partial_f A).$$

3.8. For any function $f \geqslant 0$ on Ω the set $[\hat{R}_f < R_f]$ is semipolar, and hence so is $A \cap [\hat{R}_u^A < u]$ for any $u \in \mathcal{U}$ and any set $A \subset \Omega$ (see Brelot [12, Théorème 1], Bauer [1, Satz 3.3.4]).

When applied to a strict potential u (§3.4) this shows that $i(A) = A \setminus b(A)$ is semipolar for any set A.

Any fine Borel set differs at most by a semipolar set from some Borel set in the initial topology (Constantinescu [21, Cor. 1.4]).

3.9. The fine topology on our strong harmonic space Ω is completely regular, but generally neither normal nor Lindelöf. 9) As a substitute for this latter shortcoming, Doob [23] discovered the validity of the following weaker property (see also Constantinescu [21, Cor. 1.5] for the present case of a strong harmonic space):

The quasi Lindelöf principle: Every family of finely open sets contains a countable subfamily whose union differs from that of the whole family at most by some semipolar set.

More generally: Every family of finely l.s.c. numerical

---

9) The fine topology on Ω is normal (or just as well Lindelöf) if and only if every semipolar set is countable (Berg [2]). In particular, the fine topology on a Green space is not normal.

functions defined in Ω (or in some finely open subset of Ω) contains a countable subfamily whose pointwise supremum differs from that of the given family at most in some semipolar set.

In either case, if the given family of sets or functions is upward directed, then the subsequence may be taken to be increasing.

3.10. For the theory of fine harmonicity we shall further need the property of "quasi normality" for the fine topology given in the theorem below. The proof of this theorem is based on the following simple lemma which likewise supplies a property of quasi normality (not quite a consequence of the theorem in general).

Lemma. Let A_0 and A_1 be finely closed sets such that $b(A_0) \cap b(A_1) = \emptyset$. There exists a finely continuous function f on Ω such that $f = 0$ on $b(A_0)$, $f = 1$ on $b(A_1)$, and $0 < f < 1$ elsewhere in Ω.

Proof. Let p denote a strict potential (again in the sense of Constantinescu), and put $f = f_0/(f_0 + f_1)$, where

$$f_i = p - \hat{R}_p^{A_i} \qquad (i = 0, 1). \blacksquare$$

Theorem. (The principle of quasi normality.) Let f denote a finely continuous, numerical function defined on a finely closed set $A \subset \Omega$. There exists a finely continuous function g on Ω such that $g = f$ on $b(A)$ and (if f is not constant on $b(A)$)

$$\inf_{x \in b(A)} f(x) < g < \sup_{x \in b(A)} f(x) \qquad \text{in } \complement b(A).$$

Proof. Like in the proof of the Tietze-Urysohn theorem for a normal space (see e.g. Bourbaki [6, Chap. 9, §4]) it suffices here to establish, for any finely continuous function f on A such that $|f| \leq 1$ on $b(A)$, the existence of a finely continuous function $g : \Omega \to [-1/3, 1/3]$ such that

$$|f - g| \leq 2/3 \quad \text{on} \quad b(A) ; \qquad |g| < 1/3 \quad \text{in} \quad \complement b(A) .$$

The sets

$$A_0 := \{ x \in A \mid f(x) \leq -\tfrac{1}{3} \} , \qquad A_1 := \{ x \in A \mid f(x) \geq \tfrac{1}{3} \}$$

are finely closed and disjoint, hence so are $b(A_0) \subset A_0$ and $b(A_1) \subset A_1$. By the lemma there exists a finely continuous function $g : \Omega \to [-1/3, 1/3]$ such that $g = -1/3$ on $b(A_0)$, $g = +1/3$ on $b(A_1)$, and $|g| < 1/3$ elsewhere, in particular off $b(A)$. The inequality $|f - g| \leq 2/3$ holds in $b(A_0) \cup b(A_1)$ and also in $b(A) \setminus (A_0 \cup A_1)$. To establish the same inequality in the remaining part $b(A) \cap (i(A_0) \cup i(A_1))$ of $b(A)$, note that, e.g., every point x of $b(A) \cap i(A_1)$ is such that $\{x\}$ is thin at x, and x is finely isolated in A_1 (where $f \geq 1/3$), but not in A (cf. end of §3.5), and consequently $f(x) = 1/3$ by the fine continuity of f on A. ▯

CHAPTER II

Capacity in axiomatic potential theory

4. The domination axiom

Throughout the rest of the present paper we consider a strong harmonic space Ω satisfying the domination axiom:

(D) Every locally bounded potential on Ω has the domination property (§ 1.10).

We refer to Boboc and Cornea [5] as to various equivalent forms of the domination axiom (D).

A connected, strong harmonic space satisfying axiom (D) is the same as a harmonic space in case A_1 of Brelot's axiomatic theory (including axiom (D)) as exposed in [8], [13], [16], and [29]. This identification was established by Köhn and Sieveking [30] who proved that any strong harmonic space with axiom (D) is elliptic in the sense of Bauer [1, § 5] and hence satisfies Brelot's convergence axiom in each component.

The domination axiom is a local property (for a strong harmonic space). In particular, any open subset of a strong harmonic space with axiom (D) is a space of the same kind (with respect to the restricted sheaf of harmonic functions), cf. Brelot [8, p. 131], Boboc and Cornea [5, Théorème 2].

We proceed to list some well-known consequences of the hypothesis, to be made everywhere in the sequel, that Ω is a strong harmonic space satisfying the domination axiom (D).

4.1. The only hyperharmonic function in a subdomain (= connected open subset) of Ω which is not superharmonic is the constant $+\infty$. If Ω is connected we thus have

$$\mathcal{U} = \mathcal{S} \cup \{+\infty\}.$$

4.2. The following properties of a set $E \subset \Omega$ are known to be equivalent (Brelot [8, Cor. to Theorem 31, p. 142]:

(i) E is polar (that is, $E \subset [s = +\infty]$ for some $s \in \mathcal{S}$).

(ii) $E \cap b(E) = \emptyset$ (E is thin at each of its points).

(iii) $b(E) = \emptyset$ (E is everywhere thin).

(iv) E is semipolar (the union of a sequence of everywhere thin sets).

It follows that every polar set E is finely closed, and

$$\partial_f E = i(E) = E.$$

In particular, a point $x \in \Omega$ is polar (in the sense that $\{x\}$ is polar) if and only if $\{x\}$ is thin at x. Hence a set A is thin at a point x belonging to A (in other terms $x \in i(A)$) if and only if x is a finely isolated, polar point of A (cf. § 3.5 and § 3.6).

4.3. The essence of § 4.2 may be expressed as follows: For any set $A \subset \Omega$ the set $i(A) = A \setminus b(A)$ is polar (Brelot [8, Theorem 31, p. 141]). It follows that the base operation is <u>idempotent</u>:

$$b(b(A)) = b(A).$$

In fact, $b(A) = b(\tilde{A}) = b(b(A) \cup i(A)) = b(b(A))$.

Any set B of the form $B = b(A)$ for some $A \subset \Omega$ is called a <u>base</u>. Thus B is a base if and only if $b(B) = B$, or equivalently if B is finely closed and without finely isolated, polar points.

A finely open set U will be called <u>regular</u> if $\complement U$ is a base. (This is an extension of the usual notion of a regular open set for the initial topology.) See also §12.5.

Any finely open set U is contained in the regular finely open set $\complement b(\complement U) = U \cup i(\complement U)$ $(\subset \tilde{U})$. In view of the regularity of the fine topology this shows that the regular finely open sets form a basis for the fine topology.

4.4. The term <u>quasi everywhere</u> (q.e.) will mean: everywhere except possibly in some polar set.

For any numerical function $f \geq 0$ on Ω we have (in the sense of pointwise infimum relatively to $\mathcal{F}^+$ (cf. §1.1 and §1.8)

$$\hat{R}_f = \inf \{ u \in \mathcal{U} \mid u \geq f \text{ q.e.} \}. \tag{9}$$

From the identity between polar and semipolar sets follows moreover in view of §3.8 the famous <u>convergence theorem</u> of Brelot [8, p. 133 ff.]:

$$\hat{R}_f = R_f \quad \text{quasi everywhere.}$$

In other terms: For any family of functions u_i of class $\mathcal{U}$:

$$\bigwedge_i u_i \;=\; \inf_i u_i \qquad \text{quasi everywhere.} \tag{10}$$

For any $u \in \mathcal{U}$ and any set A it follows from (9) and §3.7 that

$$\hat{R}_u^A \;=\; \hat{R}_u^{\tilde{A}} \;=\; \hat{R}_u^{b(A)} \;=\; R_u^{b(A)} \quad (= u \text{ in } b(A)).$$

Thus $\hat{R}_u^A$ is the pointwise smallest hyperharmonic function $\geqslant 0$ majorizing u on $b(A)$. By duality we obtain, moreover,

$$\mu^A = \mu^{\tilde{A}} = \mu^{b(A)} \tag{11}$$

for any admissible measure μ and any set A . Generally, it follows from (9) that the swept-out of a measure or a function on a set A does not change when A is modified by a polar set.

4.5. Lemma. For any numerical function $f \geqslant 0$ on Ω

$$R_f(x) = \max\{f(x),\ \hat{R}_f(x)\} \qquad \text{for all } x \in \Omega.$$

Proof. Write $\hat{R}_f = u \ (\in \mathcal{U})$. Then $E := [u < f]$ is polar by the convergence theorem. For any given $x_0 \in \Omega$ and $\varepsilon > 0$ it is well known that there exists $v \in \mathcal{S}$ such that $v = +\infty$ in $E \setminus \{x_0\}$ and $v(x_0) < \varepsilon$. Hence

$$u + v \in \mathcal{U}\,; \qquad u + v \geqslant f \text{ in } \complement\{x_0\}.$$

If $f(x_0) \leqslant u(x_0)$ then $u + v \geqslant f$ everywhere, and hence

$$R_f(x_0) \leqslant u(x_0) + v(x_0) \leqslant u(x_0) + \varepsilon.$$

If instead $f(x_0) > u(x_0)$, choose a finite potential w so that $w(x_0) = f(x_0) - u(x_0)$, and note that $u + v + w \geqslant f$. It follows then that

$$R_f(x_0) \leqslant u(x_0) + v(x_0) + w(x_0) \leqslant f(x_0) + \varepsilon. \ \square$$

4.6. Theorem. For any increasing sequence of numerical functions $f_n \geqslant 0$ on Ω

$$R_{\sup f_n} = \sup_n R_{f_n} \quad ; \qquad \hat{R}_{\sup f_n} = \sup_n \hat{R}_{f_n} \; .$$

Proof. Write $\hat{R}_{f_n} = u_n$, and put $\sup_n f_n = f$, $\sup_n u_n = u$. Then $u_n, u \in \mathcal{U}$, $u \geqslant u_n \geqslant f_n$ q.e. by (9), and hence $u \geqslant f$ q.e., showing that $u \geqslant \hat{R}_f$. The inequality $\hat{R}_f \geqslant u$ is obvious. Finally, in view of the above lemma,

$$R_f(x) = \max\{f(x), u(x)\} = \sup_n \max\{f_n(x), u_n(x)\} =$$

$$= \sup_n R_{f_n}(x)$$

for every $x \in \Omega$. ▯

4.7. We refer to Mme Hervé [29, §28] for the following important properties of the swept-out μ^A of an admissible measure μ on arbitrary sets $A \subset \Omega$: [10)]

a) $\mu^A = \mu^{b(A)}$ is carried by $b(A)$.

b) $[\mu^A = \mu] \Longleftrightarrow [\mu$ is carried by $b(A)]$.

c) The implication $[\mu(E) = 0] \Longrightarrow [\mu^A(E) = 0]$ holds for every polar set E, and likewise for any subset E of the fine interior of A.

---

10) The present case of an arbitrary admissible measure is easily reduced to the case considered by Mme Hervé, viz. that of a measure of compact support (see Fuglede [27]).

4.8. As a consequence of a) and c) above note that, if μ is carried by $b(\complement A)$, then so is μ^A, and μ^A is then carried by

$$b(A) \cap b(\complement A) \subset \partial_f A$$

(the fine boundary of A). Actually, we shall prove below (Lemma 12.3) that $b(A) \cap b(\complement A) = b(\partial_f A)$, and so $\mu^A = \mu^{\partial_f A}$ when μ is carried by $b(\complement A)$.

4.9. If a set A is thin at a point $x \in \Omega$, that is, if $x \in \complement b(A)$, then ε_x^A does not charge the polar sets.[11] On the other hand, if $x \in b(A)$ then $\varepsilon_x^A = \varepsilon_x$.

It follows that, for any set $A \subset \Omega$ and any point $x \in \Omega$, every numerical function f on Ω which coincides quasi everywhere with some ε_x^A-measurable function, is itself ε_x^A-measurable.

In particular, every function which is finely l.s.c. quasi everywhere (or equivalently: equal q.e. to some finely l.s.c. function g) is ε_x^A-measurable for any A and x. (In fact, each set $[g \leq \lambda]$ is finely closed, and hence differs at most by some polar set from its base which is a G_δ.)

We emphasize the case $A = \complement V$, $\mu = \varepsilon_x$, with V finely open and $x \in V$. The "harmonic measure" $\varepsilon_x^{\complement V}$ is then carried by $\partial_f V$ ($\subset \tilde{V}$) and does not charge the polar sets.

---

11) In fact, $\varepsilon_x^A(\{x\}) = 0$ according to a), and $\varepsilon_x^A(E) = 0$ according to c), §4.7, for any polar set E not containing x.

4.10. Definition. For any set $A \subset \Omega$ and any numerical function f defined in a set $U \subset \Omega$ the relation

$$f^A(x) = \int f \, d\varepsilon_x^A$$

defines a numerical function f^A on the set of all points $x \in \Omega$ such that the integral on the right exists in the wide sense.[12)]

For any A and f it follows from (11), § 4.4, that

$$f^A = f^{\tilde{A}} = f^{b(A)}.$$

Moreover, f^A is always defined at any point x of $b(A)$ at which f itself is defined; for $\varepsilon_x^A = \varepsilon_x$. Thus

$$f^A(x) = f(x) \quad \text{for every} \quad x \in b(A) \cap U.$$

Note also that f^A is everywhere defined in Ω if for instance f is a Baire function $\geqslant 0$ on Ω, or a function $\geqslant 0$ on Ω which is finely l.s.c. (or finely u.s.c.) quasi everywhere in Ω.

---

12) This means, firstly, that f should be defined almost everywhere with respect to the measure μ in question (in casu ε_x^A), in other terms, $\complement U$ should be μ-negligible; next that f should be μ-measurable, and finally that the definition

$$\int f \, d\mu = \int f^+ d\mu - \int f^- d\mu$$

should make sense, i.e., the right hand member should not be of the form $(+\infty) - (+\infty)$.

4.11. For any two functions f_1, f_2 defined in Ω and such that f_1^A, f_2^A are likewise everywhere defined,

$$[f_1 \leq f_2 \quad \text{q.e.}] \text{ implies } [f_1^A \leq f_2^A \text{ q.e.}],$$

and even $f_1^A \leq f_2^A$ everywhere off $b(A)$. In fact, $f_1^A = f_1$, $f_2^A = f_2$ in $b(A)$, and ε_x^A does not charge the polar sets when $x \in \complement b(A)$ (§4.9).

4.12. For any <u>hyperharmonic</u> function $u \geq 0$ on Ω (that is, for $u \in \mathcal{U}$) it follows from (6), §3.2, that u^A is everywhere defined (for any set A), and that

$$u^A = \hat{R}_u^A \qquad (\leq u).$$

4.13. Let μ denote an admissible measure on Ω, and A a subset of Ω. It can be shown that the family of measures $(\varepsilon_x^A)_{x \in \Omega}$ is μ-<u>adequate</u> in the sense of Bourbaki [7, Ch. 5, §3] with the integral

$$\mu^A = \int \varepsilon_x^A \, d\mu(x).$$

For any numerical function $f \geq 0$ on Ω which is μ^A-measurable we therefore have

$$\int f \, d\mu^A = \int f^A \, d\mu,$$

whereby $f^A(x) = \int f \, d\varepsilon_x^A$ is well defined for μ-almost every $x \in \Omega$. The above results will not be used in the sequel. They depend on the validity of the above identity for the particular case $f \in \mathscr{C}_0^+$, and this can be established in much the same way as Lemma 9.3 below.

5. The set functions $\hat{R}_1^{\bullet}$, $\int \hat{R}_1^{\bullet} dm$, and $R_1^{\bullet}(x_0)$

5.1. To fix the terminology, consider first an arbitrary set function C defined on all subsets of a topological space Ω and taking values in $[0, +\infty]$. Following [25] we shall call C a capacity on Ω if C is increasing, countably subadditive:

$$C(\bigcup_n A_n) \leq \sum_n C(A_n),$$

and if $C(\emptyset) = 0$.

Two sets A_1, A_2 are called C-equivalent if $C(A_1 \Delta A_2) = 0$, Δ denoting symmetric difference. Two functions f_1, f_2 on Ω are called C-equivalent if $C([f_1 \neq f_2]) = 0$.

A set $A \subset \Omega$ will be called quasi closed (with respect to C) if

$$\inf \{ C(A \Delta F) \mid F \text{ closed} \} = 0. \tag{12}$$

We shall call A quasi closed in the narrow sense if

$$\inf \{ C(A \setminus F) \mid F \text{ closed}, F \subset A \} = 0. \tag{13}$$

While the notion of a quasi closed set A as defined above in (12) only depends on the C-equivalence class of A, this is not quite generally the case for the corresponding notion in the narrow sense (13) (cf. however below).

A numerical function f on Ω is called quasi continuous (with respect to C) if

$$\inf \{ C(\omega) \mid f \text{ continuous rel. to } [\omega \} = 0 .$$

(Continuity is not understood to imply finiteness.) Quasi u.s.c. and quasi l.s.c. functions are defined similarly. These 3 notions depend only on the C-equivalence class of f .

A capacity C is called an <u>outer capacity</u> if it is continuous from the right at any set A in the following sense:

$$C(A) = \inf \{ C(G) \mid G \text{ open}, \ G \supset A \}.$$

In the case of an outer capacity C the properties (12) and (13) above are equivalent, that is, every quasi closed set is quasi closed in the narrow sense ([25, Lemma 2.2]).[13)] Similarly, the set ω in the above definitions of quasi continuity and quasi semicontinuity may be taken to be <u>open</u> when C is an outer capacity.

When Ω is locally compact and countable at infinity, the above "quasi topological" concepts are of a <u>local</u> character. For instance, a set $A \subset \Omega$ is quasi closed (resp. quasi closed in the narrow sense) if and only if $A \cap K$ has the same property for every compact set K . This follows easily by use of a countable, locally finite covering of Ω by compact sets K_n. Approximating each $A \cap K_n$ sufficiently well by closed sets F_n contained in K_n we may use the set $F = \bigcup F_n$ which is likewise closed.

---

13) Brelot uses in [14], [16], and [17] the restricted form (13) as definition of quasi closed sets. This slight variation in terminology is immaterial since the important capacities (or "weights") are mostly outer capacities.

After this digression we return to the case of a strong harmonic space satisfying the domination axiom (D).

5.2. Let m denote a (positive Radon) measure on Ω. Since every $u \in \mathcal{U}$ is l.s.c. and either everywhere $= 0$ or everywhere > 0 in any component of Ω, $\int u\,dm = 0$ implies $u = 0$ in every component of Ω charged by m.

For any downward directed family of functions $u_\alpha \in \mathcal{U}$ such that $\int u_\alpha\,dm < +\infty$,

$$\int \left(\bigwedge_\alpha u_\alpha\right) dm = \inf_\alpha \int u_\alpha\,dm$$

provided that m does not charge the polar sets. (This follows from the fact that every family (u_α) on $\mathcal{U}$ contains a countable subfamily (u_{α_n}) with the same pointwise infimum relatively to $\mathcal{U}$ (§1.1), and further that, by the convergence theorem, (10), §4.4, this infimum $\bigwedge_\alpha u_\alpha = \bigwedge_n u_{\alpha_n}$ differs from the pointwise infimum $\inf_n u_{\alpha_n}$ relatively to $\mathcal{F}^+$ only in a polar, hence m-negligible set. When (u_α) is downward directed the subsequence (u_{α_n}) may of course be taken to be decreasing.)

5.3. Each of the set functions

$$A \mapsto \hat{R}_1^A, \quad A \mapsto \int \hat{R}_1^A\,dm, \quad A \mapsto R_1^A(x_0)$$

is a <u>capacity</u> in the above sense (taken over from [25]), x_0 being any given point of Ω, and $m \geq 0$ a given measure on Ω.

The first capacity, $\hat{R}_1^{\bullet}$, takes values in $\mathcal{U}$ (cf. in this connection [25, §1.6]), the others in $[0, +\infty]$. It is clear how the notions recapitulated in §5.1 shall be interpreted in the case of a $\mathcal{U}$-valued capacity, the infimum, inf, being replaced now by $\wedge$, that is, the infimum in the complete lattice $\mathcal{U}$ (with the induced pointwise ordering). With this understanding the content of §5.1 carries over to the case of a $\mathcal{U}$-valued capacity, as is easily verified by application notably of [3 , Lemma 3.5] .

Each of the above three capacities is <u>sequentially order continuous from below</u> on account of Theorem 4.6. They take the value 0 precisely at the polar sets (not containing x_0 in the case of $R_1^{\cdot}(x_0)$).[14] This follows from §4.4 (and Lemma 4.5).

The set functions $\hat{R}_1^{\cdot}$ and $R_1^{\cdot}(x_0)$ are known to be <u>outer capacities</u>:

$$\hat{R}_1^A = \wedge \{ \hat{R}_1^G \mid G \text{ open}, \ G \supset A \},$$

$$R_1^A(x_0) = \inf \{ R_1^G(x_0) \mid G \text{ open}, \ G \supset A \}.$$

The remaining set function, $\int \hat{R}_1^{\cdot} dm$, is an outer capacity if m does not charge the polar sets and if moreover $\int \hat{R}_1 \, dm < +\infty$. This follows at once by use of § 5.2.

5.4. A fundamental result of Brelot [14] - the axiomatic version of a theorem of Choquet [20, Théorème 1] - asserts that the capacity $R_1^{\cdot}(x_0)$ is of <u>Choquet's type</u> in the sense that every finely closed set $A \subset \Omega$ is quasi closed in the narrow sense (13):

---

14) At this point we assume, in the case of $R_1^{\cdot}(x_0)$, that Ω is connected; and in the case of $\int \hat{R}_1^{\cdot} dm$ that m charges every component of Ω .

$$\inf \{ R_1^{A \setminus F}(x_0) \mid F \text{ closed}, \quad F \subset A \} = 0 .$$

Since $R_1^{\cdot}(x_0)$ is an <u>outer</u> capacity, this is the same as saying that every finely closed set is quasi closed. 15)

It follows at once that the $\mathcal{U}$-valued capacity $\hat{R}_1^{\cdot}$ likewise is of Choquet's type: For any finely closed set

$$\bigwedge \{ \hat{R}_1^{A \setminus F} \mid F \text{ closed}, \quad F \subset A \} = 0$$

(because $\bigwedge \leqslant \inf$). This implies in view of § 5.2 that the remaining capacity $\int \hat{R}_1^{\cdot} \, dm$ is of Choquet's type, at least if $\int \hat{R}_1 \, dm < +\infty$ and m does not charge the polar sets.

In view of the final remark in § 5.1 the Choquet property is a local one. Hence it suffices, in the case of $\int \hat{R}_1^{\cdot} \, dm$, to assume that $\int \hat{R}_1^K dm < +\infty$ for every compact K, that is, m is admissible in the sense of Def. 3.1. The hypothesis that m should not charge the polar sets can be dropped, as will be shown below (Theorem 15.2).

5.5. The Choquet property is known to imply the following results concerning numerical functions (in the case of any one of our 3 capacities):

Every finely continuous (resp. finely u.s.c., or finely l.s.c.) function f on Ω is quasi continuous (resp. quasi u.s.c., quasi l.s.c.).

---

15) Brelot considers a connected space Ω . This slight restriction is easily removed by passage to the harmonic subspace formed by the component of x_0 in Ω .

It even suffices to assume that f be finely continuous, etc., <u>relatively</u> to the complement of some set of capacity 0 . We refer to Brelot [14], [16], [17], and Fuglede [25, §4.3] .

5.6. Conversely, for each of the 3 capacities, C , every quasi closed set A is C-equivalent to its fine closure $\tilde{A}$, that is, $C(\tilde{A} \setminus A) = 0$.

Any quasi continuous (resp. quasi u.s.c., or quasi l.s.c.) function is finely continuous (resp. finely u.s.c., finely l.s.c.) in the complement of some set of capacity 0 .

These are well known, simple consequences of the fact that $R_1^{\bullet}(x_0)$, and hence also the other two capacities, is <u>finely stable</u>, that is, takes the same value at a set A and at its fine closure $\tilde{A}$ (as observed in § 3.7). We refer to Brelot [14], [16], [17], and Fuglede [25, §4.3, §5.5].

5.7. <u>Theorem</u> (essentially due to Brelot). The fine topology is compatible with the quasi topology determined by any one of the above 3 capacities $\hat{R}_1^{\bullet}$, $\int \hat{R}_1^{\bullet} \, dm$, and $R_1^{\bullet}(x_0)$. In the case of $\int \hat{R}_1^{\bullet} \, dm$ it is supposed that m is an admissible measure which does not charge the polar sets, but does charge every component of Ω . In the case of $R_1^{\bullet}(x_0)$ it is supposed that Ω is connected.

This compatibility, defined in [25, Def. 4.3], means that the following 2 conditions are satisfyed by the capacity C in question and by the associated quasi topological notions, C-equivalence of sets, and notion of base:

(T_1) A set $A \subset \Omega$ is $\mathcal{C}$-quasi closed if and only if it is $\mathcal{C}$-equivalent to some finely closed set.

(T_2) For any set $E \subset \Omega$ we have the biimplications [16)]

$$[\mathcal{C}(E) = 0] \Leftrightarrow [E \cap b(E) = \emptyset] \Leftrightarrow [b(E) = \emptyset].$$

Proof. Condition (T_1) is satisfied according to § 5.4 and § 5.6. As to (T_2), recall that the base $b(A)$, as defined in § 3, consists of all points $x \in \tilde{A}$ such that x is not a polar and finely isolated point of A (see also end of § 4.2). In the case of the capacities $\hat{R}_1^{\bullet}$ and $\int \hat{R}_1^{\bullet} dm$ the sets of capacity 0 are precisely the polar sets, as noted in § 5.3, and hence the above base operation b is the relevant one for the compatibility in question, as discussed in [26, § 4, § 5], in particular for the above condition (T_2), which thus is a consequence of § 4.2.

In the remaining case of the capacity $R_1^{\bullet}(x_0)$, which vanishes precisely at the polar sets not containing x_0, the relevant base operation, which we shall now denote by b_0, is obtained by viewing x_0 as a non polar point even if it is polar in the usual sense. Explicitly,

$$b_0(A) = \begin{cases} b(A) \cup \{x_0\} & \text{if } x_0 \in A, \\ b(A) & \text{otherwise}. \end{cases}$$

It is immediately verified that (T_2) holds again, now with b replaced by b_0 and $\mathcal{C}$ by $R_1^{\bullet}(x_0)$. ▯

---

16) In the case of $R_1^{\bullet}(x_0)$, b should be replaced by b_0 as explained below.

5.8. We proceed to state some known consequences of Theorem 5.7, invoking the results of [25, §4.4] applied to the $\mathcal{U}$-valued capacity $\hat{R}_1^{\cdot}$, 17) In the formulation of these results the term polar set is understood in the usual sense (as a set contained in $[s = +\infty]$ for some superharmonic function $s \geq 0$ on Ω). Likewise the term quasi everywhere (q.e.) continues to mean: everywhere except possibly in some polar set.

One of these consequences of Theorem 5.7 is Doob's quasi Lindelöf principle (quoted in §3.9 above for the more general case of an arbitrary strong harmonic space). A related result is the following concerning the lattice of bases.

5.9. (Cf. Doob [23].) The class $\mathcal{B}$ of all bases $B \subset \Omega$ (that is, sets such that $b(B) = B$) is a complete lattice under inclusion. The infimum $\bigwedge B_i$ in $\mathcal{B}$ of any family (B_i) of bases is the base of their intersection, and there is always a countable subfamily (B_{i_n}) with the same infimum:

$$\bigwedge_i B_i = b(\bigcap_i B_i) = \bigwedge_n B_{i_n} .$$

Moreover, this infimum differs from the intersection $\bigcap B_i$ (or from $\bigcap B_{i_n}$) at most by a polar set (in view of §4.3).

---

17) If one prefers to work with capacities with numerical values one may equally well use the capacity $\int \hat{R}_1^{\cdot} dm$, say with m = any harmonic measure (the restriction to a connected space being easily removed afterwards by separate application to each of the countably many components).

6. The functionals $\hat{R}_\cdot$, $\int \hat{R}_\cdot \, dm$, and $R_\cdot(x_0)$

6.1. The following convex cones of functions on Ω will enter in the subsequent discussions:

$\mathcal{F}^+$ = the cone of all numerical functions $\geqslant 0$ on Ω ,

$\mathcal{G} = \{ f \in \mathcal{F}^+ \mid f$ is l.s.c.$\}$,

$\mathcal{H}_0 = \{ f \in \mathcal{F}^+ \mid f$ is u.s.c., finite, and of compact support$\}$,

$\mathcal{C}_0^+ = \mathcal{G} \cap \mathcal{H}_0$

$= \{ f \in \mathcal{F}^+ \mid f$ is continuous, finite, and of compact support$\}$.

6.2. We shall consider the following 3 functionals on $\mathcal{F}^+$:

$$f \longmapsto \hat{R}_f \ , \quad f \longmapsto \int \hat{R}_f \, dm \ , \quad f \longmapsto R_f(x_0) ,$$

where $x_0 \in \Omega$ and the measure $m \geqslant 0$ on Ω are given. When applied to indicator functions these 3 functionals reduce to the set functions considered in the preceding section.

Each of the above functionals is clearly a <u>capacity</u> in the sense of [26 , Chap. I] , except that the first functional $\hat{R}_\cdot$ takes values in the lattice cone $\mathcal{U}$ rather than $[0, +\infty]$.

Thus each of the 3 functionals $\hat{R}_\cdot$, $\int \hat{R}_\cdot \, dm$, and $R_\cdot(x_0)$ is increasing, positive homogeneous, and countably subadditive:

$$\hat{R}_{\Sigma f_n} \leqslant \sum \hat{R}_{f_n}$$

for any sequence of functions $f_n \in \mathcal{F}^+$ (and similarly for the two other functionals).

Each of the 3 capacities is, moreover, sequentially order continuous from below (by Theorem 4.6). Each of them takes the value 0 precisely at the functions vanishing outside some polar set (not containing x_0 in the case of the capacity $R_{\cdot}(x_0)$).[18] This follows from §4.4 and Lemma 4.5.

6.3. The functionals $\hat{R}_{\cdot}$ and $R_{\cdot}(x_0)$ are upper capacities, that is, continuous from above in the sense of [26, §3]:

$$\hat{R}_f = \bigwedge \{ \hat{R}_g \mid g \in \mathcal{G},\ g \geq f \},$$

$$R_f(x_0) = \inf \{ R_g(x_0) \mid g \in \mathcal{G},\ g \geq f \}$$

for every $f \in \mathcal{F}^+$. In fact, these relations would hold even if $\mathcal{G}$ were replaced by the smaller cone $\mathcal{U}$, by the definition of $\hat{R}_f$ and R_f (§1.8).

The remaining functional, $\int \hat{R}_{\cdot}\, dm$, is an upper capacity if for instance $\int s\, dm < +\infty$ for every $s \in \mathcal{S}$ (thus in particular if m is any harmonic measure).[19] This follows at once from §5.2 since m as stated does not charge the polar sets.

---

18) At this point it is assumed, in the case of $R_{\cdot}(x_0)$, that Ω is connected; and in the case of $\int \hat{R}_{\cdot} dm$ that m charges every component of Ω.

19) A necessary and sufficient condition for $\int \hat{R}_{\cdot} dm$ to be an upper capacity is that there exists, for every polar set $E \subset \Omega$, a function $s \in \mathcal{S}$ equal to $+\infty$ in E and such that $\int s\, dm < +\infty$. This condition implies that m does not charge the polar sets.

6.4. Lemma. Let p denote a potential on Ω.

a) p is semibounded if and only if

$$\bigwedge_{\lambda} \hat{R}_{(p-\lambda)^+} = 0 .$$

b) If p is semibounded and if $\int p\, dm < +\infty$, then

$$\inf_{\lambda} \int \hat{R}_{(p-\lambda)^+}\, dm = 0 .$$

The converse implication holds, e.g., if m has compact support and charges every component of Ω. [19 bis)]

c) If p is semibounded and if $p(x_0) < +\infty$, then

$$\inf_{\lambda} R_{(p-\lambda)^+}(x_0) = 0 .$$

The converse implication holds if Ω is connected. [19 bis)]

Remarks. 1) Since $(p-\lambda)^+$ is l.s.c., we get from §3.7

$$\hat{R}_{(p-\lambda)^+} = R_{(p-\lambda)^+} .$$

Note that, in b), m is allowed to charge polar sets. Hence c) follows from b) applied to the Dirac measure at x_0. The remaining proof of a) and b) will be given in §6.7 below.

2) The lemma would remain valid if $\hat{R}_{(p-\lambda)^+} = R_{(p-\lambda)^+}$ were replaced by $\hat{R}_p^{[p>\lambda]} = R_p^{[p>\lambda]}$ throughout, but a) would then reduce to the very definition of semiboundedness given in §2.1. This will appear from the proof of b) given below.

---

19 bis) The converse implication in b) and in c) will not be used in the sequel.

6.5. Theorem. Let $f \in \mathcal{F}^+$, $p \in \mathcal{P}^s$, and $f \leq p$.

a) If f is finely u.s.c. quasi everywhere then

$$\wedge \{ \hat{R}_{f-\varphi} \mid \varphi \in \mathcal{H}_0 ,\ \varphi \leq f \} = 0.$$

If f is finely continuous quasi everywhere then

$$\wedge \{ \hat{R}_{|f-\varphi|} \mid \varphi \in \mathcal{C}_0^+ \} = 0.$$

b) Similarly with $\hat{R}_{\cdot}$ replaced by $\int \hat{R}_{\cdot}\, dm$ (and $\wedge$ by inf), provided that $\int p\, dm < +\infty$ and that m does not charge the polar sets.

c) Similarly with $\hat{R}_{\cdot}$ replaced by $R_{\cdot}(x_0)$ (and $\wedge$ by inf), provided that $p(x_0) < +\infty$ and that x_0 is one of the points at which f is finely u.s.c. (resp. finely continuous).

Remark. For any function $f \in \mathcal{F}^+$ which is finely l.s.c. quasi everywhere we have

$$\wedge \{ \hat{R}_{g-f} \mid g \in \mathcal{G},\ g \geq f \} = 0.$$

The analogous statement concerning $R_{\cdot}(x_0)$ holds when f is furthermore finely l.s.c. at the point x_0. In the case of $\int \hat{R}_{\cdot}\, dm$ it suffices to assume that m be admissible and that $\int \hat{R}_{\cdot}\, dm$ be an upper capacity (cf. §6.3). These results follow immediately from [26, Theorem 2.5 and Lemma 3.2] in view of §5.4 and §5.5 above.

Alternatively, in the latter case of $\int \hat{R}_{\cdot}\, dm$, the hypothesis that m be admissible and that $\int \hat{R}_{\cdot}\, dm$ be an upper capacity may be replaced by the weaker assumption that m does not charge the polar sets; but then the function f is required to have a l.s.c. majorant p which should be m-integrable. (This restriction allows us to confine the attention to functions $g \leq p$, and the result is obtained from the stated result concerning the capacity $\hat{R}_{\cdot}$ by integration with respect to m, cf. §5.2.

The above result concerning approximation from above enables us to reduce the proof of the stated approximation property of a finely continuous function $f (\leq p)$, in each of the 3 cases a), b), and c) of Theorem 6.5, to the asserted approximation property from below for a function f finely u.s.c. (q.e.). In this way we even obtain the slightly sharper result

$$\bigwedge \{ \hat{R}_{|f-\varphi|} \mid \varphi \in \mathcal{C}_0^+, \ \varphi \leq p \} = 0,$$

and similarly in cases b) and c). In fact,

$$\bigwedge \{ \hat{R}_{g-h} \mid g \in \mathcal{G}, \ h \in \mathcal{H}_0, \ h \leq f \leq g \leq p \} = 0$$

because $\hat{R}_{g-h} \leq \hat{R}_{g-f} + \hat{R}_{f-h}$. According to [26, Lemma 3.4] there exists $\varphi \in \mathcal{C}_0^+$ with $h \leq \varphi \leq g$, hence $|f - \varphi| \leq g - h$, $\varphi \leq p$, and consequently we arrive at the stated sharper result. The proof of the corresponding result in case c) of the capacity $R_{\cdot}(x_0)$ is quite analogous. In case b) we obtain from case a) by integration (since $\hat{R}_{|f-\varphi|} \leq p$)

$$\inf \{ \int \hat{R}_{|f-\varphi|} \, dm \mid \varphi \in \mathcal{C}_0^+, \ \varphi \leq p \} = 0$$

provided that $\int p \, dm < +\infty$, and that m does not charge the polar sets.

6.6. <u>Theorem</u> (essentially due to Brelot [15], [16]). Let (f_α) denote a downward directed family of finely u.s.c. functions $f_\alpha \in \mathcal{F}^+$ majorized by a semibounded potential p. Then

a) $$\hat{R}_{\inf f_\alpha} = \bigwedge_\alpha \hat{R}_{f_\alpha} ,$$

b) $$\int \hat{R}_{\inf f_\alpha} dm = \inf_\alpha \int \hat{R}_{f_\alpha} dm ,$$

c) $$R_{\inf f_\alpha}(x_0) = \inf_\alpha R_{f_\alpha}(x_0),$$

provided that, in c), $p(x_0) < +\infty$, and that, in b), $\int p\,dm < +\infty$ and m does not charge the polar sets.

6.7. <u>Proof</u> of Lemma 6.4 and Theorems 6.5 and 6.6. The proof will be carried out in several steps. Let us remark first that, in either theorem, b) follows from a) by integration in view of § 5.2 because only functions $\leq p$ enter effectively (cf., at this point, the remark to Theorem 6.5). On the other hand, a) in each theorem will be derived from b) applied to harmonic measures, see 3) below. At the same time Lemma 6.4 will be involved (and proved).

1) Proof of b) in Theorem 6.5 in the case of a harmonic measure m and a potential p (a priori not required semibounded) such that

$$\inf_\lambda \int \hat{R}_{(p-\lambda)^+} dm = 0. \tag{14}$$

According to §6.3 the functional $\int \hat{R}_{\cdot}\, dm$ on $\mathcal{F}^+$ is then an upper capacity in the sense of [26, §3.5]. By §5.5, p is quasi continuous with respect to the associated set function $\int \hat{R}^{\cdot}_{1}\, dm$. Since p is a potential,

$$\inf_K \int \hat{R}_p^{\complement K}\, dm = \int (\bigwedge_K \hat{R}_p^{\complement K})\, dm = 0$$

as K ranges over the compact subsets of Ω. (This follows from §1.9 and §5.2 because $\int p\, dm < +\infty$, m being a harmonic measure.) Combining this relation with (14), we infer from [26, Cor. 2.5] that the quasi continuous function p is of class $\mathcal{H}_0^*$ with respect to the upper capacity $\int \hat{R}_{\cdot}\, dm$ (cf. [26, §2.1]).

In the former statement in b) of Theorem 6.5, f is quasi u.s.c. (again by §5.5) and majorized by $p \in \mathcal{H}_0^*$. Invoking [26, Theorem 2.5], we conclude that $f \in \mathcal{H}_0^*$, and [26, Lemma 3.2] is applicable.

In the latter assertion, f is even quasi continuous, hence of class $\mathcal{G}^* \cap \mathcal{H}_0^*$ (see [26, §2.1]) according to [26, Theorem 2.5], whence the desired result by [26, Theorem 3.3]. For an alternative proof, see Remark 6.5.

2) Proof of b) in Theorem 6.6 under the same modified hypotheses on m and p as above in 1). The upper capacity $\int \hat{R}_{\cdot}\, dm$ is locally finite, that is, finite at functions of class $\mathcal{C}_0^+$. (For such a function φ it was in fact noted in §1.9 that $\hat{R}_\varphi \in \mathcal{S}_0^c$.) Since p is of class $\mathcal{H}_0^*$, as shown above in 1), the result now follows from [26, Theorem 3.6 (c)] in view of the quasi Lindelöf principle, which allows us to reduce

the assertion to the case of a decreasing sequence (f_n) .

3) Proof of a) in both theorems, now under the (apparently weaker) hypothesis

$$\bigwedge_\lambda \hat{R}_{(p-\lambda)^+} = 0$$

concerning the potential p . On account of § 5.2, this reduces immediately to the integrated forms established in 1) and 2), considering that (14) is fulfilled, again by § 5.2. Thus, in each of the two statements in a) of Theorem 6.5 (but with p as above), the left hand side $u \in \mathcal{P}$ has the property that $\int u\,dm = 0$ for every harmonic measure m , hence $u = 0$. And in a) of Theorem 6.6 (with p as above), we obtain from b)

$$\int \bigwedge_\alpha \hat{R}_{f_\alpha}\,dm \leqslant \inf_\alpha \int \hat{R}_{f_\alpha}\,dm = \int \hat{R}_{\inf f_\alpha}\,dm$$

for all harmonic measures m , whence the non trivial part of the result.

4) Proof of c) in both theorems in the case of a potential p such that $p(x_0) < +\infty$ and

$$\inf_\lambda R_{(p-\lambda)^+}(x_0) = 0 .$$

This works exactly like in the version of b) treated in 1) and 2) above, now using the upper capacity $R_\bullet(x_0)$ instead. Again §5.5 applies, and the results of [26] are used as before, observing that the potential p satisfies, by §1.9,

$$\inf_K R_p^{[K}(x_0) = 0 \ .$$

5) Proof of Lemma 6.4 a), and hence of parts a) and b) in both theorems. The "only if" part of Lemma 6.4 a) is obvious because

$$(p-\lambda)^+ \leq p \cdot 1_{[p>\lambda]} \ ,$$

and hence

$$\hat{R}_{(p-\lambda)^+} \leq \hat{R}_p^{[p>\lambda]} \ . \tag{15}$$

The "if part" of Lemma 6.4 a) follows from the result concerning a) of Theorem 6.6 established in 3), applied to the finely u.s.c. functions $p \cdot 1_{[p \geq \lambda]}$ $(\leq p)$ which decrease pointwise to the limit 0 quasi everywhere as $\lambda \to +\infty$. It follows that p is semibounded because

$$\bigwedge_\lambda \hat{R}_p^{[p>\lambda]} = \bigwedge_\lambda \hat{R}_p^{[p \geq \lambda]} = \hat{R}_p^{[p=+\infty]} = 0.$$

From the trivial "only if part" of the lemma, as established above, follows part a) of both theorems by virtue of the results obtained above in 3). And it was pointed out at the outset of the present subsection that a) implies b) in either theorem.

6) Proof of Lemma 6.4 b), and hence of part c) of the lemma and of both theorems. Suppose first that p is semibounded and that $\int p\,dm < +\infty$. From (5) in the proof of Theorem 2.2 we get by integration, replacing λ by $\lambda/2$:

$$\int \hat{R}_p^{[p>\lambda]} dm \leq \int \hat{R}_q^{[q>\lambda/2]} dm + 2\int p\,dm - 2\int q\,dm$$

for any $q \in \mathcal{P}^b$, $q \preccurlyeq p$. The first term on the right decreases to 0 as $\lambda \to +\infty$ through an increasing sequence because the integrand is m-integrable (being $\leq q \leq p$) and decreases pointwise everywhere to the limit 0 by virtue of (4), §2.1, applied to $q \in \mathcal{P}^b$.

Clearly $\int p\,dm = \sup \int q\,dm$ as q ranges over the specifically (hence also pointwise) upper directed family of all $q \in \mathcal{P}^b$ such that $q \preccurlyeq p$. By Theorem 2.2, p is in fact the specific (hence by §1.3 also the pointwise) supremum of this family, even within $\mathcal{F}^+$, and each q is l.s.c. Consequently,

$$\inf_\lambda \int \hat{R}_p^{[p>\lambda]} dm = 0 \,,$$

and a fortiori, by (15),

$$\inf_\lambda \int \hat{R}_{(p-\lambda)^+} dm = 0 \,.$$

Conversely, suppose this latter relation holds, and that m has compact support and charges every component of Ω. Then

$$u := \bigwedge_\lambda \hat{R}_{(p-\lambda)^+} \qquad (\in \mathcal{U})$$

vanishes identically on account of §5.2 because $\int u\,dm = 0$. According to a) of the lemma, this means that p is semibounded. To prove that $\int p\,dm < +\infty$, observe that, for $\lambda > 0$ and K compact,

$$p \leq \lambda \cdot 1_K + p \cdot 1_{\complement K} + (p-\lambda)^+ ,$$

$$\int p\,dm = \int \hat{R}_p\,dm \leq \lambda \int \hat{R}_1^K dm + \int \hat{R}_p^{\complement K} dm + \int \hat{R}_{(p-\lambda)^+} dm.$$

The first term on the right of this inequality is finite since $\hat{R}_1^K \in \mathcal{P}^b$ and m has compact support. The second term is finite when the interior of K contains the support of m because $\hat{R}_p^{\complement K}$ is harmonic in the interior of K. The last term is finite (and small) for large enough λ by assumption.

From part b) of the lemma follows part c) by the first remark to the lemma (§6.4). And this establishes c) in either theorem by virtue of the results obtained in 4). - This completes the proof of Lemma 6.4 and of Theorems 6.5 and 6.6. ▯

6.7. Lemma. For any semibounded potential p and any point $x_0 \in \Omega$ such that $p(x_0) < +\infty$ the capacity $R_p^{\cdot}(x_0)$ is of Choquet's type. Explicitly, for any finely closed set A,

$$\inf \{ R_p^{A \setminus F}(x_0) \mid F \text{ closed, } F \subset A \} = 0 .$$

Proof. For any $\lambda > 0$,

$$R_p^{A \setminus F}(x_0) \leq \lambda R_1^{A \setminus F}(x_0) + R_{(p-\lambda)^+}(x_0) .$$

For given $\varepsilon > 0$ choose first λ large enough so that $R_{(p-\lambda)^+}(x_0) < \varepsilon/2$, using Lemma 6.4 c), and next a closed set $F \subset A$ so that $R_1^{A \setminus F}(x_0) < \varepsilon/(2\lambda)$, using the Choquet property for $R_1^{\cdot}(x_0)$ (see §5.4). [20]

---

20) The assumption $p(x_0) < +\infty$ cannot be removed (see §15.4). On the other hand, the semiboundedness of p may be dropped e.g. in the case of a Green space. See [14] for x_0 non polar, and §15.3 below for x_0 polar.

6.8. Lemma. Every semibounded potential $p \in \mathcal{P}^s$ has the domination property. Thus, if p is harmonic off some closed set S, then $R_p^S = \hat{R}_p^S = p$. Explicitly,

$$[u \in \mathcal{U},\quad u \geq p \text{ q.e. in } S] \Longrightarrow [u \geq p].$$

Proof. Any $q \in \mathcal{P}^b$ such that $q \preccurlyeq p$ is likewise harmonic off S. Clearly $u \geq q$ q.e. in S, and so $u \geq \hat{R}_q^S$ everywhere in view of (9), §4.4. According to the domination axiom (D), q has the domination property (see §1.10), and hence $\hat{R}_q^S = q$. It follows that $u \geq q$, and we conclude from Theorem 2.2 that

$$u \geq \sup\{q \in \mathcal{P}^b \mid q \preccurlyeq p\} = p. \;\|$$

6.9. Lemma. For any polar set E there exists a semibounded potential p such that $p = +\infty$ everywhere in E.

Proof. We show, in addition, that p can be taken to be finite at prescribed points (not in E), one from each component of Ω. Clearly it suffices to consider the case where Ω is connected. In that case let $x_0 \in \complement E$ be given. Choose $q \in \mathcal{P}^b$ so that $q > 0$, and consider the outer capacity $C = R_q^{\cdot}(x_0)$. Since $x_0 \in \complement E$, we obtain in view of Lemma 4.5

$$C(E) = R_q^E(x_0) = \hat{R}_q^E(x_0) = 0.$$

Hence there are open sets $E_n \supset E$ such that $C(E_n) < 2^{-n}$. Writing $\hat{R}_q^{E_n} = q_n$, we have $q_n \in \mathcal{P}^b$ because $q_n \leq q$. Moreover, $q_n(x_0) \leq C(E_n) < 2^{-n}$. Being finite at x_0, the sum $p := \sum q_n$ is superharmonic, and indeed a potential by [22 bis, Prop. 2.2.2]. According to Theorem 2.2, p is semibounded. Clearly, $p = +\infty$ in E (where $q_n = q > 0$). $\|$

<u>Remark</u>. In view of this lemma the hypothesis $f \leq p$ in Theorem 6.5 can be weakened to $f \leq p$ <u>quasi</u> <u>everywhere</u>. (Just modify p by adding a semibounded potential equal to $+\infty$ in the polar set $[f > p]$.) A similar observation applies to Theorem 6.6 and to other results in the sequel.

7. Balayage on intersections of finely closed sets

7.1. Lemma. Let $p \in \mathcal{P}^{s}$ denote a semibounded potential, and let (B_i) denote a family of bases such that $p^{B_i} = p$ for every i. Then

$$p^{\wedge B_i} = p .$$

Proof. Recall that $u^A = \hat{R}_u^A$ for any $u \in \mathcal{U}$ and any set A (§4.12). Moreover, $\wedge B_i$ denotes the infimum of the family (B_i) in the complete lattice of all bases, ordered by inclusion (§5.9). This infimum is given by

$$\bigwedge_i B_i = b(\bigcap_i B_i) ,$$

and hence the conclusion of the lemma could be written equally well

$$p^{\cap B_i} = p$$

(cf. end of §4.4). Moreover, it suffices to consider the case of a countable family (B_i) since, in any case, there is a countable subsequence with the same infimum (§5.9). But in that case we could replace the bases B_i by general finely closed sets A_i, thus obtaining the following equivalent version of the lemma:

For any sequence (A_i) of finely closed sets such that $p^{A_i} = p$ for every i, we have

$$p^{\cap A_i} = p .$$

To prove the lemma in this form, consider closed sets $F_i \subset A_i$, and write

$$q = \bigwedge_i p^{F_i} \quad (\leq p),$$

$$r = \bigvee_i p^{A_i \setminus F_i} \quad (\leq \sum_i p^{A_i \setminus F_i}).$$

Each p^{F_i} is harmonic off F_i, and hence so is q since $q \preccurlyeq p^{F_i}$. Thus q is harmonic off $F := \bigcap F_i$, and consequently $q = q^F$ according to Lemma 6.8. Now apply Constantinescu [21, Theorem 1.1]:

$$p = p^{A_i} \preccurlyeq p^{F_i} + p^{A_i \setminus F_i} \preccurlyeq p^{F_i} + r,$$

$$p \preccurlyeq q + r = q^F + r \leq p^F + r \leq p^A + r.$$

For any $x \in \Omega$ such that $p(x) < +\infty$ it now follows that $p(x) \leq p^A(x)$ because each F_i may be so chosen that $p^{A_i \setminus F_i}(x)$ is as small as we please by virtue of Lemma 6.7. Thus $p \leq p^A$ holds quasi everywhere, hence actually everywhere (e.g. by the fine continuity of p and $p^A = \hat{R}_p^A$), and consequently $p = p^A$. ▮

7.2. <u>Lemma</u>. Let $p \in \mathcal{P}^s$, and let (A_n) be a sequence of finely closed sets. Write $p_1 = p$, and define inductively

$$p_{n+1} = (\dots((p_n^{A_1})^{A_2})\dots)^{A_{n+1}}.$$

Then

$$p^{\bigcap A_n} = \bigwedge p_n .$$

<u>Proof</u>. Writing $A = \bigcap A_n$, we have $p^A \leq p_n$ for all n because $p^A \leq p^{A_i}$ for all i, and $p^A = (p^A)^A = \dots$.

Writing $q = \bigwedge p_n$, we thus have $p^A \leq q$ $(\leq p)$. To establish the remaining inequality $q \leq p^A$ we propose to show that

$$q^{A_i} = q \qquad \text{for all } i .$$

It will then follow from the second version of the preceding lemma that $q = q^A \leq p^A$.

Now p_n decreases pointwise to the limit $\inf_n p_n$, which equals $\bigwedge p_n = q$ quasi everywhere by the convergence theorem (§4.4). For fixed i the sequence $((p_n)^{A_i})$ therefore decreases pointwise q.e. to the limit q^{A_i} because $\varepsilon_x^{A_i}$ does not charge the polar sets if $x \in \complement b(A_i)$, while $\varepsilon_x^{A_i} = \varepsilon_x$ if $x \in b(A_i)$ (§4.9). Consequently

$$q^{A_i} = \bigwedge_n (p_n)^{A_i} .$$

Clearly,

$$(p_n)^{A_i} \geq (\ldots(((p_n)^{A_1})^{A_2})^{A_3}\ldots)^{A_{n+1}} = p_{n+1}$$

for all n such that $n+1 \geq i$. It follows that

$$q^{A_i} = \bigwedge_n (p_n)^{A_i} = \bigwedge_{n \geq i-1} (p_n)^{A_i} \geq \bigwedge_{n \geq i-1} p_{n+1} = q ,$$

and so indeed $q^{A_i} = q$ for every i. ▮

The following theorem, which is based on Lemma 7.2, will serve in the next chapter as the key to the theory of fine harmonicity, in establishing the fine boundary minimum principle (Theorem 9.1).

On the other hand, Lemma 7.1 above implies that $\mathcal{P}^\wedge \cap Q = \{0\}$, where Q denotes the class of all $s \in \mathcal{S}$ (see §1.4) such that s is finely harmonic in $[s < +\infty]$ (or equivalently: off some polar set). Actually, Q is the band orthogonal to $\mathcal{P}^\wedge$ in $\mathcal{S}$ (§11.16).

7.3. Theorem. Let f denote a numerical function $\geqslant 0$ which is finely u.s.c. quasi everywhere in Ω and majorized in Ω by some semibounded potential. The family of all bases B such that $f \leqslant f^B$ q.e. is then stable under infimum in the lattice of all bases.

Equivalent formulation of the conclusion: For any sequence (A_i) of finely closed sets such that $f \leqslant f^{A_i}$ q.e. we have $f \leqslant f^{\cap A_i}$ q.e.

(Recall the definition $f^A(x) = \int f \, d\varepsilon_x^A$ for any set $A \subset \Omega$ and any point $x \in \Omega$ (§4.10).)

Proof (in the second formulation, cf. §7.1). Fix a function $h \in \mathcal{H}_0$ such that $h \leqslant f$ (cf. §6.1). Next consider any $\varphi \in \mathcal{C}_0^+$ such that $\varphi \geqslant h$, and any finite and continuous potential $r > 0$. By the approximation theorem of Mme Hervé [29, Théorème 6.1] there exist $p, q \in \mathcal{P}^c$ such that $p - q \in \mathcal{C}_0^+$ and

$$|p - q - \varphi| \leqslant r. \tag{16}$$

It follows that $h \leqslant \varphi \leqslant (p-q) + r$, and hence

$$f \leqslant h + R_{f-h} \overset{\text{q.e.}}{\leqslant} (p-q) + r + \hat{R}_{f-h}. \tag{17}$$

Extending slightly Definition 4.10 in the case of a function $f \geqslant 0$ on Ω (that is, $f \in \mathcal{F}^+$), we write, in this proof,

$$f^A(x) = \int^* f \, d\varepsilon_x^A$$

for any set A and any point x. Thus $f \longmapsto f^A$ defines a pointwise increasing mapping of $\mathcal{F}^+$ into itself. Moreover, as in §4.11, $f \leqslant g$ q.e. implies $f^A \leqslant g^A$ q.e.

Returning to the given function $f \in \mathcal{F}^+$, finely u.s.c. quasi everywhere in Ω , and the given finely closed sets A_i such that $f \leqslant f^{A_i}$ q.e., we define by repeated application of the above operation

$$f_n = (\ldots(((\ldots(((((f^{A_1})^{A_2})^{A_1})^{A_2})^{A_3})\ldots)^{A_1})^{A_2})\ldots)^{A_n}$$

for $n \geqslant 2$. Similarly p_n , q_n , and $(p-q)_n = p_n - q_n$ are derived from p , q , and $p - q$ ($\geqslant 0$), (cf. also §7.2). It follows that $f \leqslant f_n$ q.e. From (17) we thus obtain in view of §4.12

$$f \leqslant f_n \leqslant (p_n - q_n) + r + \hat{R}_{f-h} \qquad \text{q.e.} \tag{18}$$

Writing $A = \bigcap A_i$, we get from Lemma 7.2 combined with §1.1 and the convergence theorem (§4.4)

$$p^A = \bigwedge_n p_n = \widehat{\inf_n p_n} \overset{\text{q.e.}}{=} \inf_n p_n ,$$

and similarly for q^A. Since the sequences (p_n) and (q_n) are pointwise decreasing, it follows that

$$p^A - q^A = \lim_n (p_n - q_n) \qquad \text{pointwise q.e.} \tag{19}$$

From (16) follows

$$p^A - q^A = (p-q)^A \leqslant \varphi^A + r . \tag{20}$$

In the limit as $n \to \infty$, (18) yields together with (19) and (20)

$$f \leqslant \varphi^A + 2r + \hat{R}_{f-h} \qquad \text{q.e.} \tag{21}$$

Since Ω is locally compact with countable base, the function $h \in \mathcal{H}_0$ is representable as the pointwise infimum of a decreasing sequence of functions $\varphi_j \in \mathcal{C}_0^+$. Replacing φ by φ_j , and r by r/j in the above analysis, we infer from (21), letting $j \to \infty$,

$$f \overset{\text{q.e.}}{\leqslant} h^A + \hat{R}_{f-h} \leqslant f^A + \hat{R}_{f-h}$$

because $h \leqslant f$. Finally it follows from Theorem 6.5 a), combined with a remark in §1.1, that there exists a sequence (h_n) of functions $h_n \in \mathcal{H}_0$, $h_n \leqslant f$, such that

$$0 = \bigwedge_n \hat{R}_{f-h_n} \overset{\text{q.e.}}{=} \inf_n \hat{R}_{f-h_n}.$$

Consequently, $f \leqslant f^A$ quasi everywhere. ▯

CHAPTER III

Finely harmonic and finely hyperharmonic functions

As in the preceding chapter we consider a strong harmonic space Ω satisfying the domination axiom (D).

8. Definitions and examples

8.1. Definition. A numerical function f, defined in a finely open set $U \subset \Omega$, is called finely hyperharmonic (in U) if f is finely l.s.c. and $> -\infty$ in U, and if the induced fine topology on U has a basis consisting of finely open sets V of fine closure $\tilde{V} \subset U$ such that

$$f(x) \geqslant \int^* f \, d\varepsilon_x^{\complement V} \qquad \text{for every } x \in V.$$

Recall that $\varepsilon_x^{\complement V}$ is then carried by the fine boundary $\partial_f V \subset U$ and does not charge the polar sets (§4.9). - As to the notions of upper and lower integral of numerical, signed functions see Bourbaki [7, Chap. IV, §4, exerc. 5, 6].

8.2. Definition. A numerical function f on U as above is called finely hypoharmonic if $-f$ is finely hyperharmonic. Explicitly, f should be finely u.s.c. and $< +\infty$, and for sets V as above we should have

$$f(x) \leqslant \int_* f \, d\varepsilon_x^{\complement V} \qquad \text{for every } x \in V.$$

8.3. Definition. A real valued function f, defined in a finely open set $U \subset \Omega$, is called finely harmonic (in U) if f is finely continuous in U, and if the fine topology on U has a basis consisting of finely open sets V with $\tilde{V} \subset U$ such that f is integrable with respect to $\varepsilon_x^{\complement V}$ for every $x \in V$ and

$$f(x) = \int f \, d\varepsilon_x^{\complement V} \qquad \text{for every } x \in V.$$

Clearly, every finely harmonic function in U is both finely hyperharmonic and finely hypoharmonic in U. The converse is likewise true, but far from obvious. See Cor. 1 to Lemma 9.5 below.

8.4. Definition. For any numerical function f defined in a given finely open set U we denote by $\mathcal{D}(f)$ the class of all finely open sets V of compact closure $\bar{V}$ (in the initial topology) contained in U, and such that f is bounded on $\bar{V}$.

In the sequel we shall often use the fact that there exists, for any $V \in \mathcal{D}(f^-)$, a potential p of class $\mathcal{P}^c$ such that $f \geq -p$ in $\tilde{V}$ $(\subset \bar{V})$. (Choose $q \in \mathcal{P}^c$, $q > 0$, and note that f^-/q is bounded on $\bar{V}$, say $\leq k$, and so $f^- \leq p := kq$ on $\bar{V}$.)

If f is finely locally bounded, say from below, then $\mathcal{D}(f^-)$ is a basis for the fine topology on U. (This is because every fine neighbourhood of a point of U contains a compact one.)

More generally, for any basis $\mathcal{D}$ for the fine topology on U, those sets V which belong to both $\mathcal{D}$ and $\mathcal{D}(f^-)$, form a basis for the fine topology on U.

Accordingly, it would cause no change in the notion of fine hyperharmonicity if the finely open sets V occurring in Def. 8.1 were required to be of class $\mathcal{D}(f^-)$.

Likewise it would cause no change if the upper integral in Def. 8.1 were replaced by the corresponding lower integral. (For any $V \in \mathcal{D}(f^-)$ we have, in fact, $\int^* f \, d\varepsilon_x^{\complement V} = \int_* f \, d\varepsilon_x^{\complement V}$ because f is finely l.s.c. and hence $\varepsilon_x^{\complement V}$-measurable (§4.9).

Similar observations apply to the notion of, say, fine harmonicity, with the obvious changes such as replacing $\mathcal{W}(f^-)$ by $\mathcal{W}(f)$.

8.5. It will follow from Lemma 9.5 below that, for every finely hyperharmonic function f in U and every $V \in \mathcal{W}(f^-)$,

$$f(x) \geq \int^* f \, d\varepsilon_x^{\complement V} \qquad \text{for every } x \in U.$$

Hence, in Def. 8.1, one might equally well write "for every $x \in U$" in place of "for every $x \in V$".

This remark may be amplified as follows: If f is finely hyperharmonic in U, and if V denotes <u>any</u> finely open set with $\tilde{V} \subset U$ such that $f(x) \geq \int_* f \, d\varepsilon_x^{\complement V}$ for every (or just quasi every) $x \in V$, then the same inequality holds for every x in all of U.

In fact, extending f arbitrarily to Ω, and applying Theorem 9.13 below to the base $B := b(\complement V)$ and the function $-f$, we find that $u(x) := \int_* f \, d\varepsilon_x^{\complement V}$ is finely hyperharmonic in the finely open set $\{x \in \complement B \mid u(x) > -\infty\}$. Since moreover f is finely continuous in U by Theorem 9.10 below, the inequality $f \geq u$ extends by fine continuity from q.e. in V (hence q.e. in $\complement B$) to all of $\complement B$, and is trivial in $B \cap U$.

Finally, note that, since the regular finely open sets form a basis for the fine topology (§4.3), it would suffice to consider <u>regular</u> finely open sets V (even of class $\mathcal{W}(f^-)$) in Def. 8.1.

Again similar observations apply, say, to fine harmonicity.

<u>Problem</u>. The following question arises by comparison with the theory of ordinary hyperharmonic functions: Let f be finely l.s.c. and $> -\infty$ in U (finely open), and suppose that each point $x \in U$ has a fundamental system of fine neighbourhoods V with $\tilde{V} \subset U$ such that $f(x) \geq \int^* f \, d\varepsilon_x^{\complement V}$. Is then f finely hyperharmonic in U?

One might even assume that this fundamental system consists of all sufficiently small sets V (with $x \in V$) taken from a prescribed basis for the fine topology on U, e.g. the basis of all regular finely open sets of class $\mathcal{D}(f^-)$.

Similarly in the case of fine harmonicity. These problems are open even in the classical case of a Green space.

8.6. In view of the regularity of the fine topology, the finely harmonic functions form a sheaf (more precisely: a complete presheaf), and so do the finely hyperharmonic functions.

Explicitly: A function f defined in the union of a family of finely open sets U_i is finely harmonic (resp. finely hyperharmonic) in this union if and only if, for every i, the restriction of f to U_i is finely harmonic (resp. finely hyperharmonic) in U_i.

8.7. Theorem. In an open set $U \subset \Omega$ (in the initial topology) every hyperharmonic function is finely hyperharmonic, and every harmonic function is finely harmonic.

Proof. Let u be hyperharmonic in U. For any given point $x_0 \in U$ let U_0 denote an open neighbourhood of x_0 with compact closure $\overline{U}_0 \subset U$. By the extension theorem of Mme Hervé [29, Théorème 13.1] there exist potentials p, q on Ω such that q is harmonic in U_0 and $u = p - q$ in U_0. Consider any usual regular open set ω such that $x_0 \in \omega \subset \overline{\omega} \subset U_0$. For any finely open set $V \subset \omega$ and any $x \in V$ we have

$$p(x) \geq \hat{R}_p^{\complement V}(x) = \int p \, d\varepsilon_x^{\complement V},$$

$$q(x) = \int q \, d\varepsilon_x^{\complement \omega} = \hat{R}_q^{\complement \omega}(x) \leq \hat{R}_q^{\complement V}(x) = \int q \, d\varepsilon_x^{\complement V},$$

and so $q(x) = \int q \, d\varepsilon_x^{\complement V}$. Consequently, $u(x) \geq \int u \, d\varepsilon_x^{\complement V}$.

If u is even harmonic in U, then also the above p is harmonic in U_0, and so $p(x) = \int p \, d\varepsilon_x^{\complement V}$. Hence $u = p - q$ is integrable with respect to $\varepsilon_x^{\complement V}$ with the integral $u(x)$. ▯

For a result in the opposite direction see Theorem 9.8 below, and further Theorem 10.14. See also §8.11.

8.8. A numerical function u, defined quasi everywhere in some finely open set U, is said to be finely harmonic quasi everywhere in U if there is a polar set e such that u is defined and finely harmonic in the finely open set $U \setminus e$.

The following definition makes sense in view of the quasi Lindelöf principle (§3.9).

Definition. The quasi harmonic support, $S_q(u)$, of a numerical function u defined q.e. in Ω is the complement of the largest finely open set in which u is finely harmonic q.e.

Clearly, $S_q(u)$ is a base.

As an example note that every strict potential p in the sense of Constantinescu [21] has the whole space Ω as quasi harmonic support, that is, p is not finely harmonic in any non void, finely open set. This follows from [21, Cor. 1.2]. The problem arises whether, conversely, every potential p with $S_q(p) = \Omega$ is strict in the same sense. (Such a potential has the usual harmonic support $S(p) = \Omega$, and hence it is strict in the sense of Bauer, [1, p. 72].)

Every function u on Ω has a fine harmonic support, $S_f(u)$, defined as the smallest finely closed set off which u is finely harmonic. It is not always a base. See §9.16.

8.9. Similarly we may define the <u>quasi support</u>, $\operatorname{supp}_q \mu$, of a positive (or just real) measure μ on Ω as the complement of the largest among all finely open sets U such that μ is carried by the union of $\complement U$ and some polar set. (In view of the quasi Lindelöf principle the union of all such finely open sets has the same property.) - Clearly, $\operatorname{supp}_q \mu$ is a base.

If a measure μ does not charge any polar set, then $\operatorname{supp}_q \mu$ actually carries μ and hence is the smallest finely closed set carrying μ. Thus μ has a <u>fine support</u>, viz. $\operatorname{supp}_f \mu = \operatorname{supp}_q \mu$.

8.10. <u>Theorem</u>. Let μ denote a real measure on a Green space Ω, and suppose that $|\mu|$ is admissible (cf. §2.6, §3.1).

a) For any finely open set $U \subset \Omega$, $G\mu = G\mu^+ - G\mu^-$ is well defined and finely harmonic in U, if and only if $|\mu|^*(U) = 0$ and $G|\mu| < +\infty$ in U.

b) $S_q(G\mu) = \operatorname{supp}_q \mu$. The fine harmonic support of $G\mu$ is $S_f(G\mu) = \operatorname{supp}_q \mu \cup [G\mu$ is infinite$]$.

<u>Proof</u>. The "if" part of a) reduces to the case $\mu \geqslant 0$. Accordingly, suppose that $\mu \geqslant 0$ does not charge U, and that $G\mu < +\infty$ in U. For any finely open set V with $\tilde{V} \subset U$ we thus have $\mu^{\complement V} = \mu$ by §4.7 because μ is carried by $b(\complement V)$ $(\supset \complement\tilde{V} \supset \complement U)$. Hence

$$(G\mu)^{\complement V} = \hat{R}^{\complement V}_{G\mu} = G(\mu^{\complement V}) = G\mu ,$$

and so $G\mu$ is finely harmonic in U (being also finely continuous).

As to the "only if" part of a), we clearly have $G|\mu| = G\mu^+ + G\mu^- < +\infty$ in U. Choose $q \in \mathcal{P}_0^c$ so that $q \neq 0$, and hence $q > 0$ in Ω. Since U is covered by the finely open sets $U_n := U \cap [G|\mu| < nq]$, it suffices to prove that each $|\mu|^*(U_n) = 0$.

Thus we may assume from the beginning that $G|\mu| < q$ in U for some $q \in \mathcal{G}_0^c$. It follows that $G|\mu| \leq q$ in $\tilde{U}$. In view of Theorem 2.6, the trace ν of $|\mu|$ on the G_δ-set $\tilde{U} = b(U)$ therefore does not charge any polar set because

$$\int G\nu \, d\nu \leq \int G|\mu| d\nu \leq \int q \, d\nu \leq \int q \, d|\mu| < +\infty .$$

Hence $|\mu|$ itself does not charge any polar subset of $\tilde{U}$.

Now consider any finely open set V with $\tilde{V} \subset U$ such that $G\mu = (G\mu)^{\complement V}$ in V, that is (since e.g. $(G\mu^+)^{\complement V} \leq G\mu^+ < +\infty$),

$$G\mu^+ - G\mu^- = (G\mu^+)^{\complement V} - (G\mu^-)^{\complement V} \quad \text{in } V .$$

Hence

$$G\mu^+ + G[(\mu^-)^{\complement V}] = G\mu^- + G[(\mu^+)^{\complement V}]$$

in V, and trivially in $b(\complement V)$, hence q.e. in Ω, and indeed everywhere by fine continuity. It follows that

$$\mu^+ + (\mu^-)^{\complement V} = \mu^- + (\mu^+)^{\complement V},$$

showing that μ^+ and μ^- have the same trace on $\complement b(\complement V)$ since $(\mu^+)^{\complement V}$ and $(\mu^-)^{\complement V}$ are carried by $b(\complement V)$. But μ^+ and μ^- are disjoint measures, so this trace must be 0, and hence $\complement b(\complement V)$ is a null set for μ^+ and μ^-, and therefore for $|\mu|$. In particular, $|\mu|^*(V) = 0$ because $V \subset \complement b(\complement V)$.

According to Def. 3.1, the above sets V cover U. By the quasi Lindelöf principle we conclude that $|\mu|^*(U) = 0$ because $|\mu|$ does not charge any polar subset of U.

As to b), the stated expression for the quasi harmonic support follows easily from a), and the last identity hence from §9.16.

<u>Remark</u>. With the notations of the theorem it is easily shown that $G\mu$ is well defined and finely hyperharmonic in U if U is

a null set for μ^- and $G\mu^- < +\infty$ in U. Whether the converse implication holds remains an open question.

Example. We give an example of an admissible measure $\mu \geq 0$ on a Green space Ω, and a finely open set $U \subset \Omega$ such that $\mu^*(U) = 0$, and such that $G\mu(x) = +\infty$ for some $x \in U$. It is well known, in fact, that there exists a finely closed set $A \subset \Omega$, thin at some point $x \in \complement A$, and such that nevertheless $\hat{R}^A_{G\varepsilon_x}(x) = +\infty$, in other words that $\mu := \varepsilon^A_x$ has infinite energy 21)

$$\int G\mu \, d\mu = \int G\varepsilon_x \, d\mu = G\mu(x) = \hat{R}^A_{G\varepsilon_x}(x) = +\infty .$$

Then μ does not charge the polar sets (§4.9), and μ is carried by A, and yet $G\mu$ equals $+\infty$ at the point x of $U := \complement A$.

8.11. The notion of fine harmonicity is effectively more general than that of ordinary harmonicity.

Theorem. Let Ω be a Green space. On every finely open set $U \subset \Omega$ such that U is not open in the initial topology there exists a finely harmonic function u which is not the restriction to U of an ordinary harmonic (or just finely continuous and $< +\infty$) function defined in an open set $\supset U$.

Proof. Let x_0 denote a point of U not interior to U (in the initial topology). Any connected open neighbourhood ω_0 of x_0 in Ω therefore meets $\complement U$. According to [27, Théorème 2], or Cor. 9.8 below, ω_0 is finely connected and hence meets $\partial_f U$.

---

21) In $\Omega = \mathbb{R}^3$ there exists a closed set B admitting an equilibrium potential of infinite energy, see Cartan [18, p. 277]. Thus B is "thin at infinity", and we may take for A the image of B under inversion with respect to a point x not in B.

Consider any sequence of polar points $x_n \in \partial_f U$ converging to x_0 in the initial topology. If G is the Green kernel on Ω, choose a sequence (c_n) of real numbers > 0 so that

$$\sum c_n G(x_0, x_n) < +\infty ,$$

and hence $\sum c_n < +\infty$. The discrete measure μ on Ω obtained by assigning the mass c_n to the point x_n, $n = 1, 2, \ldots$, is then a Radon measure of compact support formed by the points x_n together with their limit x_0. The potential

$$G\mu = \sum c_n G(\cdot, x_n)$$

is finite at x_0 and hence finely harmonic in $\complement(\{x_n\}_{n \in \mathbb{N}})$ $(\supset U)$ by Theorem 8.10. Let u denote the restriction of $G\mu$ to U. Let ω denote any open set (in the initial topology on Ω) such that $\omega \supset U$. Then x_n is in ω for some n. Any finely continuous (or just finely u.s.c.) extension v of u from U to ω must take the value $+\infty$ at this point x_n because $x_n \in \partial_f U$. In fact, $v(x_n) \geq G\mu(x_n) = +\infty$ because $\{x \in \omega \mid v(x) \geq G\mu(x)\}$ is finely closed relatively to ω and contains U, hence also $x_n \in \partial_f U \cap \omega$.

Remark. The question arises whether (say in a Green space Ω) there exists, for every finely open set without interior points, a finely harmonic function u on U such that no restriction of u to a non void, finely open subset of U admits a harmonic extension. For any finely harmonic function u in a finely open set U there exists, by Theorems 8.10 and 9.8, a finely open, fine neighbourhood V of any given point of U such that u coincides in V with a locally bounded Green potential $G\mu$ of a real measure μ on Ω such that $|\mu|^*(V) = 0$.

9. Fundamental properties

9.1. Theorem. (The fine boundary minimum principle.) Let f be finely hyperharmonic in a finely open set $U \subset \Omega$, and suppose that

$$\operatorname*{fine\,lim\,inf}_{x \to y,\ x \in U} f(x) \geq 0 \tag{22}$$

for quasi every $y \in \partial_f U$. If moreover there exists a semibounded potential p on Ω such that $f \geq -p$ in U, then $f \geq 0$ (in U).

Proof. The function h defined in Ω by $h = f^-$ in U and $h = 0$ off U, is finely u.s.c. quasi everywhere in Ω (viz. everywhere except exactly at those points of $\partial_f U$ at which (22) fails to hold). Clearly $h \leq p$ everywhere.

In view of the quasi Lindelöf principle (cf. §3.9) there is a sequence of sets V_i, selected among the sets V occurring in Def. 8.1, and covering U up to a polar set. The inequality

$$h(x) \leq h^{\complement V_i}(x) \quad \left(:= \int h \, d\varepsilon_x^{\complement V_i}\right)$$

holds trivially for every $x \in b(\complement V_i)$ since then $\varepsilon_x^{\complement V_i} = \varepsilon_x$. According to Def. 8.1 the same inequality holds at every point $x \in V_i$ since $\varepsilon_x^{\complement V_i}$ is then carried by $\partial_f V_i \subset \tilde{V}_i \subset U$ (§4.9), and $-f \leq f^- = h$ in U, and consequently (if $h(x) > 0$)

$$h(x) = -f(x) \leq -\int^* f \, d\varepsilon_x^{\complement V_i} \leq \int h \, d\varepsilon_x^{\complement V_i}.$$

Altogether, $h \leqslant h^{\complement V_i}$ holds q.e. in Ω (viz. everywhere except possibly in the polar set $(\complement V_i) \setminus b(\complement V_i)$ (cf. §4.3). According to Theorem 7.3,

$$h \overset{\text{q.e.}}{\leqslant} h^{\bigcap \complement V_i} = h^{\complement U} = 0$$

because $h = 0$ on the finely closed set $\complement U$ carrying all $\varepsilon_x^{\complement U}$. Thus $f^- = 0$ q.e. in U, that is, $f \geqslant 0$ q.e. in U. For any $x \in U$ choose V as in Def. 8.1 with $x \in V$, and recall that $\varepsilon_x^{\complement V}$ does not charge the polar sets (§4.9). It follows that

$$f(x) \geqslant \int^* f \, d\varepsilon_x^{\complement V} \geqslant 0. \quad \blacksquare$$

Remarks. 1) The above fine boundary minimum principle contains that of Brelot [9, Lemma 1] which corresponds to the case U open and relatively compact, f hyperharmonic and bounded from below.

2) The semiboundedness of p cannot be dropped because we only require (22) to hold q.e. on the fine boundary. Example in the greenian case: $f = -p = -G\varepsilon_x$, $U = \complement\{x\}$. As to the possibility of dropping the semiboundedness of the potential p at the expense of supposing (22) to hold everywhere in $\partial_f U$ (corresponding to what is known for ordinary hyperharmonic functions) see Lemma 10.14 c), Remark 10.15, and §10.17 below. See also Theorem 10.8.

3) A probabilistic proof of Theorem 9.1 above was given by Nguyen-Xuan-Loc [33, Theorem 6].

9.2. Theorem. (The fine domination principle.) Every semibounded potential $p \in \mathcal{P}^s$ has the fine domination property:

$$\hat{R}_p^S = R_p^S = p$$

for any finely closed set $S \subset \Omega$ such that p is finely harmonic in $\complement S$.

Proof. For every finely open set V with $\tilde{V} \subset \complement S$ such that $p = p^{\complement V}$ in V we obtain, writing $u := \hat{R}_p^S$,

$$u - p \geq (u-p)^{\complement V} \quad \text{in} \quad V$$

because $u \geq u^{\complement V}$ $(= \hat{R}_u^{\complement V})$, and $p^{\complement V} \leq p < +\infty$ in U. Hence $u - p$ is finely hyperharmonic and $\geq -p$ in $\complement S$. Moreover, $u - p$ is defined and finely continuous q.e. in Ω, and $u - p \geq 0$ q.e. in $\partial_f(\complement S) \subset S$. Thus $u - p$ satisfies the boundary condition (22) in Theorem 9.1. ▯

Remark. It would suffice to assume concerning the finely closed set S in the above theorem that p be finely harmonic quasi everywhere in $\complement S$, that is, finely harmonic in $\complement(S \cup e)$ for some polar set e; for then $\hat{R}_p^S = \hat{R}_p^{S \cup e}$ $(= p)$. Thus it would suffice to take for S the quasi harmonic support $S_q(p)$ of p (§8.8).

9.3. Lemma. Let $p \in \mathcal{P}^\Delta$. Let f be a finely continuous numerical function defined on a base B and such that $|f| \leq p$ on B. Then f^B is defined, finite, and finely continuous in $[p < +\infty]$, and finely harmonic in $(\complement B) \cap [p < +\infty]$. Moreover, $|f^B| \leq p$ in $[p < +\infty]$, and

$$(f^B)^A = f^B \qquad \text{in } [p < +\infty] \quad \text{for any set } A \supset B.$$

Proof. The fine harmonicity of f^B follows from the other assertions when we take $A = \complement V$ in the last relation, V

being finely open and of fine closure $\tilde{V} \subset (\complement B) \cap [p < +\infty]$. And for these remaining assertions it suffices to consider the case $f \geqslant 0$. By the principle of quasi normality (Theorem 3.10) f admits a finely continuous extension to all of Ω . Since p is finely continuous in Ω , we may assume that this extension, likewise to be denoted by f , satisfies $0 \leqslant f \leqslant p$ in Ω .

Fix a potential $q \in \mathcal{P}^c$ such that $q > 0$. For any given function $\varphi \in \mathcal{C}_0^+$, and any given number $\varepsilon > 0$, there exist according to the approximation theorem $p_1 , p_2 \in \mathcal{P}^c$ such that $p_1 - p_2 \in \mathcal{C}_0^+$ and

$$|\varphi - (p_1 - p_2)| \leqslant \varepsilon q .$$

Since $f - \varphi$ is finely continuous, $|f - \varphi| \leqslant \hat{R}_{|f-\varphi|}$ everywhere, and hence

$$-v \leqslant f - (p_1 - p_2) \leqslant v \tag{23}$$

where

$$v := \hat{R}_{|f-\varphi|} + \varepsilon q \quad (\in \mathcal{U}).$$

Writing $u = f^B$, we obtain from (23) by "balayage" (integration with repsect to ε_x^B)

$$-v \leqslant u - (p_1^B - p_2^B) \leqslant v \tag{24}$$

because $v^B = \hat{R}_v^B \leqslant v$.

Since p_1^B , p_2^B , and v are finely continuous, the inequalities (24) remain valid when u is replaced by its finely u.s.c. envelope $\tilde{u}$, or by its finely l.s.c. envelope $\underset{\sim}{u}$. It follows that

$$\tilde{u} - \underset{\sim}{u} \leqslant 2v = 2\hat{R}_{|f-\varphi|} + 2\varepsilon q \quad \text{in } [p < +\infty]$$

because $[p < +\infty] \subset [v < +\infty] \subset [\tilde{u} < +\infty]$ since $\hat{R}_f \leqslant p$

and $\hat{R}_\varphi \in \mathcal{P}_0^c$ (§1.9). This implies when $\varepsilon \to 0$ that

$$\tilde{u}(x) - \underset{\sim}{u}(x) \leq 2\hat{R}_{|f-\varphi|}(x), \qquad x \in [p<+\infty],$$

and next, by varying φ, that $\tilde{u}(x) - \underset{\sim}{u}(x) = 0$ according to Theorem 6.5 c) which states that, when $p(x) < +\infty$,

$$\inf \{ \hat{R}_{|f-\varphi|}(x) \mid \varphi \in \mathcal{C}_0^+ \} = 0$$

since $\hat{R} \leq R$. Consequently, $u = f^B$ is <u>finely continuous</u> in $[p<+\infty]$. In particular, u^A is defined on all of Ω (cf. §4.10).

For any set $A \supset B$ it follows in the same way from (24), now by integration with respect to ε_x^A, that

$$-v \leq u^A - (p_1^B - p_2^B) \leq v. \tag{25}$$

In fact, $(p_i^B)^A = p_i^B$ ($i = 1, 2$) by [27, Lemme 3.1] since $B \subset A$. Combining (24) and (25), we obtain

$$|u^A - u| \leq 2v \qquad \text{in} \quad [p < +\infty],$$

whence, as above, $u^A = u$ in $[p < +\infty]$.

9.4. <u>Theorem</u>. (The global aspect of fine hyperharmonicity.) Let U be a finely open set. Let f be a numerical function defined and finely l.s.c. in $\tilde{U}$ and finely hyperharmonic in U. If moreover $f \geq -p$ in U for some $p \in \mathcal{P}^s$ then

$$f(x) \geq \int^* f \, d\varepsilon_x^{\complement U}$$

for every $x \in U \cap [p<+\infty]$ (thus in particular q.e. in U).

Proof. Write $b(\complement U) = B$, hence $\varepsilon_x^{\complement U} = \varepsilon_x^B$ by (11), §4.4. Let g denote the finely l.s.c. envelope of the function equal to f in U and to $+\infty$ in $\complement U$. Then g is finely l.s.c. and $\geq -p$ in Ω, $g \geq f$ in $\tilde{U}$, and $g = f$ in U.

By the quasi Lindelöf principle there is an increasing sequence of finely continuous numerical functions f_n on Ω converging to g pointwise q.e. (Here we use also the complete regularity of the fine topology.)

Let q denote a fixed potential > 0 of class $\mathcal{P}^c$. Replacing, if necessary, f_n by $-p$ on $[f_n < -p]$, and by nq on $[f > nq]$, we may assume that

$$-p \leq f_n \leq nq,$$

and hence (cf. §4.10)

$$-p \leq f_n^B \leq nq.$$

For $m > n$ the function $f_{n,m}$ defined by

$$f_{n,m}(x) = \max\{f_n(x), -mq(x)\}$$

is finely continuous, and $|f_{n,m}| \leq mq$. According to the above lemma, $f_{n,m}^B$ is finely continuous in Ω, and

$$(f_{n,m}^B)^A = f_{n,m}^B \qquad \text{for every } A \supset B. \tag{26}$$

For $m \to +\infty$ we obtain $f_{n,m} \to f_n$ decreasingly, and hence $f_{n,m}^B \to f_n^B$ because $f_{n,m}^B \leq nq^B \leq nq < +\infty$. It follows that f_n^B is finely u.s.c., and in the limit from (26) that

$$(f_n^B)^A = f_n^B \qquad \text{for every } A \supset B \tag{27}$$

because $nq^A \leqslant nq < +\infty$. Now g is finely hyperharmonic in U (being $= f$ there), and hence

$$g^{\complement V} \leqslant g$$

holds in V for certain finely open sets V with $\tilde{V} \subset U$ forming a basis for the fine topology in U . For each of these sets V we infer from (27), applied to $A = \complement V$ $(\supset \complement U \supset B)$ that

$$(f_n^B)^{\complement V} = f_n^B$$

everywhere in Ω . Consequently, $g - f_n^B$ is defined, finely l.s.c., and $\geqslant -(p + nq)$ in $[g > -\infty]$ $(\supset [p < +\infty])$, and moreover finely hyperharmonic in U $(\subset [g > -\infty])$ because

$$(g - f_n^B)^{\complement V} \leqslant g - f_n^B$$

in V for each of the above sets V . Moreover,

$$g - f_n^B = g - f_n \geqslant 0 \quad \text{in} \quad B \cap [p < +\infty],$$

hence q.e. in $\partial_f U \subset \complement U$. Since $p + nq \in \mathcal{P}^A$, we conclude from the fine boundary minimum principle (Theorem 9.1) that $g - f_n^B \geqslant 0$ in U , that is,

$$g(x) \geqslant \int f_n \, d\varepsilon_x^B \quad \text{for every} \quad x \in U. \tag{28}$$

Finally, as $n \to +\infty$, f_n increases, and $f_n \to g$ q.e. Since $\varepsilon_x^B = \varepsilon_x^{\complement U}$ does not charge the polar sets when $x \in U$, and since $g \geq f$ in $\tilde{U}$, which contains $\partial_f U$ and hence carries $\varepsilon_x^{\complement U}$ (§4.9), we conclude from (28) that

$$f(x) = g(x) \geq \sup_n \int f_n \, d\varepsilon_x^B = \int g \, d\varepsilon_x^B \geq \int f \, d\varepsilon_x^{\complement U}$$

at any point $x \in U$ such that $p(x) < +\infty$. ▯

Remark. Under the hypotheses of the theorem we even have

$$f(x) \geq \int_* f \, d\varepsilon_x^{\complement U} \qquad \text{for every } x \in U.$$

(This inequality holds q.e. in U by the theorem, and extends to all of U by fine continuity as in §8.5.) - This stronger result means that $f \geq f^{\complement U}$ at any point $x \in U$ such that $\int f^- d\varepsilon_x^{\complement U} < +\infty$. (Recall that the lower integral $\int_* f$ equals $-\infty$ when $\int f^- = +\infty$, even in the indeterminate case where also $\int f^+ = +\infty$.)

Corollary. (The global aspect of fine harmonicity.) Let U be a finely open set. Let f be a numerical function defined and finely continuous in $\tilde{U}$ and finely harmonic in U . If moreover $|f| \leq p$ in U for some $p \in \mathcal{P}^s$, then

$$f(x) = \int f \, d\varepsilon_x^{\complement U}$$

for every $x \in U \cap [p < +\infty]$ (and more generally, by the above remark, for every $x \in U$ such that f is integrable with respect to $\varepsilon_x^{\complement U}$).

9.5. Lemma. Let f be finely hyperharmonic in U (finely open). Then $f \geq f^{\complement V}$ holds in V (and even in U) for any finely open set V with $\tilde{V} \subset U$ such that $f \geq -p$ in V for some finite and semibounded potential p .

In particular, $f \geq f^{\complement V}$ holds in U for any V belonging to the basis $\mathcal{W}(f^-)$ for the fine topology in U (§8.4).

Proof. For any set V as stated Theorem 9.4 is applicable with V in place of U (or with $V \cup i(\complement V)$ in place of U).

Corollary 1. A function f is finely harmonic in U if (and only if) f is both finely hyperharmonic and finely hypoharmonic in U.

In fact, we then obtain $f = f^{\complement V}$ in V for any $V \in \mathcal{W}(f)$.

Remark. The basis $\mathcal{W}(f^-)$ for the fine topology in U depends on the order of magnitude of the negative part f^- of the finely hyperharmonic function f in question (unlike the situation for ordinary hyperharmonic functions in an open set, such functions being bounded from below on every compact set).

For any family (f_i) of finely hyperharmonic functions f_i in U (finely open), having a common minorant which is finely hyperharmonic (or just finely locally bounded from below), such a minorant f allows us in view of the above lemma to specify a common basis for the fine topology in U (e.g. the basis $\mathcal{W}(f^-)$) with the property that $f_i \geqslant f_i^{\complement V}$ in V holds for every i and for every V from that basis.

This observation applies, in particular, to any finite family of finely hyperharmonic functions. - There are of course similar observations regarding finely harmonic functions.

Corollary 2. The finely hyperharmonic functions in the finely open set U form a convex subcone of $[-\infty, +\infty]^U$, stable under pointwise supremum for upper directed families, and under pointwise infimum for finite families. - The finely harmonic functions in U form a linear subspace of $\mathbb{R}^U$.

9.6. Lemma. The cone of finely hyperharmonic functions in U (finely open) is closed (within $[-\infty, +\infty]^U$) under finely locally uniform convergence, and so is the vector space of finely harmonic functions in U. 22)

Proof. Replacing, if necessary, U by a suitably small finely open fine neighbourhood of any given point of U, we may assume from the beginning that f is the uniform limit of a net of finely hyperharmonic functions in U. Then f is finely l.s.c. and $> -\infty$ in U. Further we may assume that there exists in some open set W containing U a harmonic function $h \geqslant 1$. By assumption there exists, for any $\varepsilon > 0$, a finely hyperharmonic function g in U such that $f - \varepsilon \leqslant g \leqslant f + \varepsilon$, and hence

$$f - \varepsilon h \leqslant g \leqslant f + \varepsilon h .$$

Clearly, $\mathcal{W}(f^-) = \mathcal{W}(g^-)$, and $\mathcal{W}(f^-)$ is a basis for the fine topology in U. For every $V \in \mathcal{W}(f^-)$ we have in V by Lemma 9.5, since h is finely hyperharmonic in W by Theorem 8.7,

$$f + \varepsilon h \geqslant g \geqslant g^{\complement V} \geqslant (f - \varepsilon h)^{\complement V} \geqslant f^{\complement V} - \varepsilon h,$$

and hence $f \geqslant f^{\complement V} - 2\varepsilon h$ in V. Letting $\varepsilon \to 0$, we infer that $f \geqslant f^{\complement V}$ in V for every such V, and consequently f is finely hyperharmonic in U. ▯

---

22) The uniformity in question on the range space $[-\infty, +\infty]$ is supposed here to be the one defined by the pseudo-distance $|a - b|$ between extended real numbers a and b. (We put $a - a = 0$ even if a is infinite.)

In place of the finely locally uniform convergence it suffices, in the case of finely harmonic functions, to assume finely locally <u>bounded</u> convergence (Theorem 11.9 below). The same nearly applies to the case of finely hyperharmonic functions ≥ 0 (Cor. 11.3).

9.7. <u>Lemma</u>. Let $f \geq 0$ be defined in Ω and finely <u>hypoharmonic</u> in U (finely open). The hyperharmonic function $u := \hat{R}_f$ on Ω is then finely harmonic in $U_0 := U \cap [u < +\infty]$. Moreover, $u^{\complement V} = u$ for any finely open set V with $\tilde{V} \subset U$ such that $f \leq p$ in V for some $p \in \mathcal{P}^s$.

<u>Proof</u>. For any V as stated it follows from §4.11 (since $u \geq f$ q.e.) and from Theorem 9.4 that

$$u^{\complement V} \geq f^{\complement V} \geq f \qquad \text{q.e. in } \Omega .$$

(Note that $f^{\complement V} = f$ in $b(\complement V)$, hence q.e. in $\complement V$.) Since $u^{\complement V} = \hat{R}_u^{\complement V}$ is hyperharmonic in Ω, we get from (9), §4.4, $u^{\complement V} \geq \hat{R}_f = u$, and consequently $u^{\complement V} = u$.

Taking, in particular, V so that even $\tilde{V} \subset U_0$, the result obtained implies (see §8.4) that u is finely harmonic in U_0. ∎

<u>Corollary</u>. For any <u>superharmonic</u> function $s \geq 0$ in Ω (that is, $s \in \mathcal{S}$), and any set $A \subset \Omega$, the swept-out function $s^A = \hat{R}_s^A$ (cf. §4.12) is finely harmonic in $[s^A < +\infty] \cap \complement b(A)$, and hence finely harmonic q.e. in $\complement b(A)$. The quasi harmonic support $S_q(s^A)$ of s^A (Def. 8.8) is thus contained in $b(A)$.

In the case where s is a semibounded potential we shall determine completely the quasi harmonic support $S_q(s^A)$ in terms of $S_q(s)$ in Theorem 13.4 below.

9.8. Theorem. Let u denote a numerical function defined in an open set U (for the initial topology).

a) u is hyperharmonic if and only if u is finely hyperharmonic and moreover locally bounded from below (in the initial topology).

b) u is harmonic if and only if u is finely harmonic and moreover locally bounded.

Proof. The "only if" part was established in Theorem 8.7 above. As to the "if" part of a), it follows from Lemma 9.5 that $u \geq u^{\complement V}$ in V for every usual regular open set V of compact closure $\overline{V} \subset U$ because u is supposed to be bounded from below on $\overline{V}$, and so V is of class $\mathcal{D}(u^-)$ (§8.4). Having thus shown that u is nearly hyperharmonic in U, we conclude that u is actually hyperharmonic there, being also finely l.s.c. (see Boboc, Cornea, and Constantinescu [4, Lemmas 1.1 and 2.4]). b) follows from a). ▯

As to the possibility of dropping, or relaxing, the condition of local boundedness in the "if" parts of the theorem, see §10.15 and §10.16 below.

Corollary. An open set U is finely connected if and only if U is connected (in the initial topology).

In fact, for any decomposition $U = V \cup W$ of U into disjoint finely open sets V and W, the function u on U defined by $u = +\infty$ in V and $u = 0$ in W is finely hyperharmonic in U, hence hyperharmonic there, in particular l.s.c., and so V must be open. Similarly, W is open, and consequently V or W must be void since U is connected. ▯

For a more direct proof of this corollary see [27, Théorème 2]. - Note also that the corollary can be given the following equivalent formulation:

For any open set U the fine components of U are the same as the components of U in the initial topology.

9.9. Theorem. (The local extension property.) Let f be finely hyperharmonic in U (finely open). Every point $x_0 \in U \cap [f < +\infty]$ has then a finely open fine neighbourhood V with $\tilde{V} \subset U$ such that f is representable in $\tilde{V}$ as the difference $u - v$ between two locally bounded potentials u, v (on Ω) of which v is finely harmonic in V.

Proof. Suppose first that f is bounded in some fine neighbourhood of x_0. Let $p \in \mathcal{P}^c$ be a strict potential >0 on Ω in the sense of Constantinescu [21, p. 278] (see also §3.4 above). Then $1/p$, and hence f/p, is bounded in some finely open set containing x_0. We may therefore assume from the beginning that f/p is bounded in all of U, say $|f| \leq p$ in U.

Choose a regular finely open set U_0 (§4.3) with $x_0 \in U_0 \subset \tilde{U}_0 \subset U$. Writing for brevity

$$q := p^{\complement U_0} = \hat{R}_p^{\complement U_0},$$

we have by (8), §3.4,

$$p > q \quad \text{in } U_0 ; \qquad p = q \quad \text{in } \complement U_0 . \tag{29}$$

Since $f \geq -p$ in U, it follows from Lemma 9.5 that

$$f \geq f^{\complement U_0} \geq -p^{\complement U_0} = -q \quad \text{in } U,$$

and hence

$$-q \leq f \leq p \qquad \text{in } U .$$

Since $p(x_0) > q(x_0)$ by (29), there is a constant $\lambda > 0$ such that

$$\lambda p(x_0) - \lambda q(x_0) > p(x_0) + 2q(x_0) .$$

The set

$$V := [\lambda p - \lambda q > p + 2q]$$

is finely open, and $x_0 \in V \subset \tilde{V} \subset U_0$ on account of (29) and the fine continuity of p and q.

Now define a function u on Ω by

$$u(x) = \begin{cases} \min \{ \lambda p(x), \ f(x) + (\lambda + 2) q(x) & \text{for } x \in U, \\ \lambda p(x) & \text{for } x \in \complement \tilde{U}_0 . \end{cases}$$

This definition makes sense because $U \cup \complement \tilde{U}_0 = \Omega$, and $\lambda p < f + (\lambda + 2) q$ in $U \setminus U_0$ (since $p = q$ and $f \geq -p$ there).

Clearly, u is finely hyperharmonic in U and in $\complement \tilde{U}_0$, hence in all of Ω (by the sheaf property, §8.6). Since $u \geq 0$, it follows from the preceding theorem that u is hyperharmonic, hence $u \in \mathcal{P}^b$ because $u \leq \lambda p \in \mathcal{P}^b$. Moreover, $v := (\lambda + 2) q \in \mathcal{P}^b$, and v is finely harmonic in $V \subset U_0$ by Lemma 9.3 since $q = p^{\complement U_0}$. Finally, $u = f + v$ in $\tilde{V}$ because

$$\lambda (p - q) \geq p + 2q \geq f + 2q \qquad \text{in } \tilde{V} .$$

Note that $f = u - v$ is finely continuous in $\tilde{V}$.

To complete the proof of the theorem, it remains to prove that every finely hyperharmonic function f is bounded in some fine neighbourhood of any point of $[f < +\infty]$. But this follows from the next theorem, the proof of which merely depends on the validity of the present theorem in the finely locally bounded case which we have just considered. ▯

9.10. <u>Theorem</u>. Every finely hyperharmonic function is finely continuous.

<u>Proof</u>. Let f be finely hyperharmonic in U (finely open). Then f is finely l.s.c. and $> -\infty$, in particular finely locally bounded from below. It remains to prove that f is finely u.s.c. at any point $x_0 \in U$ where $f(x_0) < +\infty$.

Let h denote a usual (hence also finely) harmonic function > 0 , defined and bounded in some open neighbourhood W of x_0, and such that $h(x_0) = 1$. For any number $\lambda > f(x_0) = f(x_0)/h(x_0)$, the function $g := \min\{f, \lambda h\}$ is finely hyperharmonic in $U \cap W$ by Cor. 9.5, and finely locally bounded there. Hence it follows from the proof given in §9.9 that g is finely continuous at every point of $U \cap W$. Consequently, the set

$$\{x \in U \cap W \mid f(x)/h(x) < \lambda\} = \{x \in U \cap W \mid g(x) < \lambda h(x)\}$$

is finely open, and it contains x_0 by the choice of λ . This shows that f/h is finely u.s.c. at x_0 , and so is therefore f . ▯

9.11. Theorem. (Monotone nets of finely harmonic functions.) The pointwise limit f of a monotone net (f_i) of finely harmonic functions in a finely open set U is finely harmonic in the finely open set $U \cap [\,|f| < +\infty\,]$.

Proof. Let the net (f_i) be increasing. By Corollary 2 to Lemma 9.5, $f := \sup f_i$ is finely hyperharmonic in U and hence finely continuous according to the preceding theorem. In particular, $U \cap [\,|f| < +\infty]$ is finely open.

By virtue of the quasi Lindelöf principle there is an increasing sequence (i_n) such that $f = \sup f_{i_n}$ quasi everywhere in U, hence actually everywhere in U because $\sup f_{i_n}$ is likewise finely hyperharmonic and hence finely continuous in U.

For any finely open set V of compact closure $\overline{V} \subset U$ such that f_{i_n} is bounded on V for all sufficiently large n, we have $f_{i_n} = f_{i_n}^{\complement V}$ in V (by Cor. 9.4 or Lemma 9.5), and hence $f = f^{\complement V}$ in V. It follows that f is finely harmonic in $U_0 := U \cap [\,|f| < +\infty]$. In fact, $f_{i_1} \leq f_{i_n} \leq f$, and here f_{i_1} and f are finely locally bounded in U_0 (being finite and finely continuous there). According to §8.4, those finely open sets V with compact closure $\overline{V} \subset U_0$ such that $|f_{i_1}| + |f|$ is bounded on $\overline{V}$, form a basis for the fine topology in U_0, and we have seen that $f = f^{\complement V}$ holds in V for any such V. $\square$

Remark. The proof shows, moreover, that every increasing net of finely hyperharmonic functions in U admits an increasing subsequence with the same pointwise supremum (likewise finely hyperharmonic).

Corollary. The fine topology on Ω is locally connected.

In fact, the proof given in Bauer [1, Satz 1.1.10] for the local connectivity of the initial topology on a harmonic space carries over mutatis mutandis, using now the fine topology and finely harmonic functions, and applying Theorem 9.10 or 9.11.

A more direct proof of this corollary was given in [27, Théorème 4].

As another application of Theorem 9.11 we proceed to extend part of Lemma 9.3 to more general functions f on the base B. For the sake of simplicity we shall suppose that the semibounded potential p in the following lemma is finite. - The results thus obtained will be used in the study of the generalized fine Dirichlet problem (§14).

9.12. Lemma. Let p denote a finite and semibounded potential, B a base, and f a finely l.s.c. numerical function defined and $\geq -p$ in B. The function $u = f^B$ defined in all of Ω by

$$u(x) = \int f \, d\varepsilon_x^B$$

is finely l.s.c. and $\geq -p$ in Ω, finely hyperharmonic in $\complement B$, and finely harmonic in the finely open set $(\complement B) \cap [u < +\infty]$. Moreover, $u^A = u$ for any set $A \supset B$.

Proof. The extension of f to all of Ω defined by $f = +\infty$ in $\complement B$, to be denoted likewise by f, is again finely l.s.c. and $\geq -p$. As noted in §4.9, f is ε_x^B-measurable for every $x \in \Omega$, and $u = f^B$ is therefore well defined and $\geq -p^B \geq -p$ in all of Ω.

Fix $q \in \mathcal{P}^c$, $q > 0$. By the complete regularity of the fine topology, f is the pointwise supremum of the family Φ of all finely continuous φ on Ω such that $-p \leq \varphi \leq f$ and such that moreover φ/q is bounded from above. Since ε_x^B does not charge the polar sets when $x \in \complement B$, it follows easily by application of the quasi Lindelöf principle that

$$u = f^B = \sup \{ \varphi^B \mid \varphi \in \Phi \}$$

(Note that this holds trivially in B since $f^B = f$, $\varphi^B = \varphi$ there.)

Now Lemma 9.3 is applicable to each $\varphi \in \Phi$, and the stated properties of $u = f^B$ follow on account of Cor. 9.5 and Theorem 9.11. In fact, as above,

$$u^A = \sup \{ (\varphi^B)^A \mid \varphi \in \Phi \} = \sup \{ \varphi^B \mid \varphi \in \Phi \} = u$$

for every set $A \supset B$. ▯

9.13. <u>Theorem</u>. Let f be any numerical function defined in a base B. The function u defined in Ω by

$$u(x) = \int^* f \, d\varepsilon_x^B$$

is finely hyperharmonic in the finely open set

$$U^- := \{ x \in \complement B \mid \int_* f^- \, d\varepsilon_x^B < +\infty \},$$

finely hypoharmonic in the finely open set

$$U^+ := \{ x \in \complement B \mid \int^* f^+ \, d\varepsilon_x^B < +\infty \}$$
$$= (\complement B) \cap [u < +\infty],$$

and hence finely harmonic in the finely open set $U^+ \cap U^-$ of all points of $\complement B$ at which u is finite. Moreover, $u = u^{\complement V}$ holds in $V \cap (U^+ \cup U^-)$ for every finely open set V with $\tilde{V} \subset \complement B$.

<u>Proof</u>. Extend f to Ω by putting, say, $f = 0$ in $\complement B$. Consider first the case where $0 \leq f \leq p$ for some finite and semibounded potential p. Then (with $\mathcal{G}$ as in §6.1)

$$u(x) = \int^* f \, d\varepsilon_x^B = \inf \{ g^B(x) \mid g \in \mathcal{G}, \ f \leq g \leq p \}$$

for every $x \in \Omega$. By the preceding lemma, each g^B is here finely harmonic in $\complement B$, and hence so is u by Theorem 9.11. Clearly, $u \leq p^B \leq p$ in Ω.

For general $f \geq 0$, choose $p \in \mathcal{P}^c$, $p > 0$, and define

$$f_n(x) = \min \{ f(x), \ np(x) \}; \qquad u_n(x) = \int^* f_n \, d\varepsilon_x^B .$$

Then each u_n is finely harmonic in $\complement B$, and Theorem 9.11 is again applicable, showing that $u = \sup u_n$ is indeed finely hyperharmonic in $\complement B$, and finely harmonic in the finely open set $(\complement B) \cap [u < +\infty]$. Moreover, since $u_n \leq np$, it follows from Cor. 9.4 that $u_n = u_n^{\complement V}$ in V, and hence $u = u^{\complement V}$ in V, for every finely open V such that $\tilde{V} \subset \complement B$.

Next consider the case $f \leq 0$. Here (with $\mathcal{H}_0$ as in §6.1)

$$- u(x) = \int_* (-f) \, d\varepsilon_x^B = \sup \{ h^B(x) \mid h \in \mathcal{H}_0, \ h \leq -f \} \tag{30}$$

for every $x \in \Omega$. Any $h \in \mathcal{H}_0$ is majorized by a potential $q \in \mathcal{P}^b$ $(\subset \mathcal{P}^s)$, see §1.9, and hence each h^B above is $\leq q$ in Ω, and finely harmonic in $\complement B$ by the preceding lemma applied to $-h$. Consequently, u is finely hypoharmonic in $\complement B$ and finely harmonic in $(\complement B) \cap [u > -\infty]$ according to Theorem 9.11. The relation $u = u^{\complement V}$ in V for V as stated reduces to the similar relation for h^B (valid by Cor. 9.4) by application of the quasi Lindelöf principle to (30), noting that $-u$ is finely l.s.c. in $\complement B$, and that each h^B is even finely continuous there.

Finally, the case of a quite arbitrary function f reduces to the cases just considered because

$$u(x) = \int^* f\, d\varepsilon_x^B = \int^* f^+ d\varepsilon_x^B + \int^* (-f^-)\, d\varepsilon_x^B$$

in $U^+ \cup U^-$, viz. for any $x \in \complement B$ where this latter difference is not of the form $(+\infty) + (-\infty)$. With $\leq$ in place of equality, this follows from Bourbaki [7, Chap. IV, §4, exerc. 5 b)]. As to the opposite inequality, we may assume that there exist l.s.c. and μ-integrable functions $g \geq f$ (where $\mu := \varepsilon_x^B$). For any such g,

$$\int g\, d\mu = \int g^+ d\mu + \int (-g^-)\, d\mu \geq \int^* f^+ d\mu + \int^* (-f^-)\, d\mu,$$

whence the result by taking infimum with respect to g. ▯

<u>Remark</u>. Consider again a numerical function f defined on a base B. Let p denote a finite and semibounded potential on Ω. If $f \geq -p$ (on B), then, for every $y \in B$,

$$\underset{x \to y,\ x \in \Omega}{\text{fine lim inf}} \int_* f\, d\varepsilon_x^B = \underset{x \to y,\ x \in B}{\text{fine lim inf}}\ f(x). \qquad (31)$$

Similarly, if $f \leq p$ (on B), then, for every $y \in B$,

$$\underset{x\to y,\ x\in\Omega}{\text{fine lim sup}} \int^* f\, d\varepsilon_x^B = \underset{x\to y,\ x\in B}{\text{fine lim sup}}\ f(x) .$$

In particular, if $|f| \leqslant p$ on B, then each of the extensions

$$x \longmapsto \int_* f\, d\varepsilon_x^B , \qquad x \longmapsto \int^* f\, d\varepsilon_x^B ,$$

of f from B to Ω is finely continuous at any point $y \in B$ at which f is finely continuous (relatively to B).

As to (31), the inequality $\leqslant$ is obvious. To establish the opposite inequality $\geqslant$, let $g : B \longrightarrow [-\infty, +\infty]$ denote the greatest finely l.s.c. minorant of f on B. Then $-p \leqslant g \leqslant f$ (on B), and

$$\int g\, d\varepsilon_x^B \leqslant \int_* f\, d\varepsilon_x^B$$

for every $x \in \Omega$, whence the result by Lemma 9.12, according to which $\int g\, d\varepsilon_x^B$ is finely l.s.c. as a function of $x \in \Omega$; for it takes the value $g(y) = \underset{x\to y,\ x\in B}{\text{fine lim inf}}\ f(x)$ at every $y \in B$.

9.14. Theorem. (Removable singularities for finely hyperharmonic functions.) Let u be finely hyperharmonic in $U \setminus e$, where e denotes a polar subset of the finely open set U. Then u admits a finely hyperharmonic extension to U if (and only if) u is bounded from below in some deleted fine neighbourhood of each point of e. The extension is then unique and given by

$$u(y) = \underset{x\to y,\ x\in U\setminus e}{\text{fine lim}}\ u(x) , \qquad y \in e .$$

Proof. Recall that e is finely closed and finely discrete. Suppose that the stated (obviously necessary) condition is fullfilled, and extend u to U as follows:

$$u(y) = \underset{x\to y,\ x\in U\setminus e}{\text{fine lim inf}}\ u(x) , \qquad y \in e .$$

This extension u is finely l.s.c. and $> -\infty$ in U. Let V be a regular, finely open set (that is, $\complement V$ is a base) with $\tilde{V} \subset U$ and such that $u \geq -p$ in V for some finite and semibounded potential p on Ω (for instance, let V belong to the basis for the fine topology in U formed by all regular sets of class $\mathcal{D}(u^-)$ (taken relatively to U), cf. §4.3 and §8.4).

Since u is finely hyperharmonic and $\geq -p$ in $V \setminus e$, and finely l.s.c. in $(V \setminus e)^{\sim} = \tilde{V} \subset U$, it follows from Theorem 9.4 that

$$u \geq u^{\complement(V \setminus e)} = u^{\complement V} \qquad \text{in } V \setminus e.$$

Define f on $\complement V$ by $f = u$ on $\partial_f V$, and $f = +\infty$ in $\complement \tilde{V}$. Then f is finely l.s.c. and $\geq -p$ in $\complement V$. According to Lemma 9.12, applied to the base $\complement V$, $f^{\complement V}$ is finely hyperharmonic in V, hence finely l.s.c. there (and even finely continuous by Theorem 9.10). For any $x \in V$, $\varepsilon_x^{\complement V}$ is carried by $\partial_f V \subset U$ (§4.9), and so $u^{\complement V}(x) = \int u \, d\varepsilon_x^{\complement V}$ is well defined and equal to $f^{\complement V}(x)$ for $x \in V$. Having thus established that $u^{\complement V}$ is defined and finely l.s.c. in V, in particular at at any point of $V \cap e$, we conclude that the inequality $u \geq u^{\complement V}$ in $V \setminus e$ extends to all of V. For any $y \in V \cap e$ we have, in fact

$$u(y) = \underset{x \to y,\ x \in V \setminus e}{\text{fine lim inf}}\ u(x) \geq \underset{x \to y,\ x \in V \setminus e}{\text{fine lim inf}}\ u^{\complement V}(x) \geq u^{\complement V}(y).$$

This shows that u is finely hyperharmonic in U, in particular finely continuous. Consequently we have the stated expression for $u(y)$, $y \in e$, whence the uniqueness of the extension. ▯

Remark. In view of Theorem 8.7, the above theorem contains as a particular case the following well-known result:

Let u be hyperharmonic and bounded from below in an open set

ω (in the initial topology). Then the fine limit of u exists at every irregular boundary point y for ω. (Just apply Theorem 9.14 to the finely open set $U = \omega \cup \{y\}$.)

9.15. Theorem. (Removable singularities for finely harmonic functions.) Let u be finely harmonic in $U \setminus e$ (with U finely open, and $e \subset U$ polar). Then u admits a finely harmonic extension to U if (and only if) u is bounded in some deleted fine neighbourhood of each point of e. The extension is unique and given by

$$u(y) = \underset{x \to y,\, x \in U \setminus e}{\text{fine lim}} u(x), \qquad y \in e .$$

Proof. By the preceding theorem, u can be extended (if the stated, obviously necessary condition is satisfied) to a finely hyperharmonic function u_1 in U, and also to a finely hypoharmonic function u_2 in U. Since u_1 and u_2 are finely continuous in U by Theorem 9.10, and coincide with u in the finely dense subset $U \setminus e$ of U, they are identical and hence yield a finely harmonic extension of u from $U \setminus e$ to U. The uniqueness is obvious. ▯

Corollary. With U and e as above, if u is finite and finely continuous in U and finely harmonic in $U \setminus e$, then u is finely harmonic in all of U.

9.16. Every numerical function u defined in all of Ω has a fine harmonic support, denoted by $S_f(u)$, and defined as the complement of the largest finely open set in which u is finely harmonic. Clearly $S_f(u)$ contains the quasi harmonic support, $S_q(u)$, defined in §8.8. The difference is polar, and hence

$$S_q(u) = b(S_f(u)).$$

If u is finely continuous off $S_q(u)$, then

$$S_f(u) = S_q(u) \cup [|u| = +\infty]$$

by the above corollary.

In particular, $S_f(u) = S_q(u)$ holds if (and only if) u is finite and finely continuous off $S_q(u)$.

In a similar way one may define the quasi harmonic support $S_q(u)$ and the fine harmonic support $S_f(u)$ relative to a given finely open set $U \subset \Omega$, u being defined in U. One uses then the induced fine topology on U, and similarly the base of a subset A of U should be understood relatively to U, that is as the intersection between U and the previously considered base of A. The above observations then remain in force (with Ω replaced by U).

10. Finely superharmonic functions and fine potentials

10.1. Lemma. Let u, resp. v, be finely hyperharmonic in a finely open set U, resp. V. Suppose that $V \subset U$ and that

$$\operatorname*{fine\,lim\,inf}_{x \to y,\ x \in V} v(x) \geq u(y) \qquad \text{for every } y \in U \cap \partial_f V.$$

Then the following function w is finely hyperharmonic in U :

$$w(x) = \begin{cases} \min\{u(x), v(x)\} & \text{for } x \in V, \\ u(x) & \text{for } x \in U \setminus V. \end{cases}$$

Proof. Choose $q \in \mathcal{P}^c$, $q > 0$. For every $\varepsilon > 0$ put

$$w_\varepsilon(x) = \begin{cases} \min\{u(x), v(x) + \varepsilon q(x) & \text{for } x \in V, \\ u(x) & \text{for } x \in U \setminus V. \end{cases}$$

Clearly w and w_ε are finely l.s.c. and $> -\infty$ in U. The set

$$A := \{x \in V \mid u(x) < v(x) + \varepsilon q(x)\}$$

is finely open since u is finely continuous (Theorem 9.10). In view of the given relative boundary inequality, every point $y \in U \cap \partial_f V$ has a fine neighbourhood $W_y \subset U$ such that $W_y \cap V \subset A$, that is, $W_y \subset A \cup (U \setminus V)$. It follows that $A \cup (U \setminus V)$ is finely open. In this subset of U, w_ε coincides with u and so is finely hyperharmonic. On the other hand, w_ε is finely hyperharmonic in V by Cor. 9.5, being the pointwise infimum of the two finely hyperharmonic functions u and $v + \varepsilon q$ there. By the sheaf property (§8.6), w_ε is finely hyperharmonic in U, and so is therefore w. In fact, w is the finely locally uniform limit of w_ε in U as $\varepsilon \to 0$ (Lemma 9.6) because

$w \leq w_\varepsilon \leq w + \varepsilon q$. $\square$

The above proof is slightly more complicated than the proof for the case of ordinary hyperharmonic functions in open sets. This is tied up with the open question raised at the end of §8.5.

10.2. <u>Theorem</u>. Let u be finely hyperharmonic in a finely open set U. Let V denote a regular finely open set such that $\tilde{V} \subset U$, and suppose that $u \geqslant -p$ in V for some finite and semibounded potential p on Ω. Then $u^{\complement V}$ (defined in U) is the smallest function which is finely hyperharmonic in U, $\geqslant u$ in $U \setminus V$, and $\geqslant -p$ in V. Moreover, $u^{\complement V}$ is finely harmonic in $V \cap [u^{\complement V} < +\infty]$ (and $u^{\complement V} \leq u$ in U, $u^{\complement V} = u$ in $U \setminus V$).

<u>Proof</u>. According to Lemma 9.5, $u^{\complement V} \leqslant u$ in V. Clearly, $u^{\complement V} = u$ in $U \cap b(\complement V) = U \setminus V$. Now apply Lemma 9.12 to the base $\complement V$ and to the finely l.s.c. function f defined in Ω by

$$f = \begin{cases} u & \text{in } \tilde{V}, \\ +\infty & \text{in } \complement \tilde{V}. \end{cases}$$

Since $u \geqslant -p$ in V (hence in $\tilde{V}$), we have $f \geqslant -p$ in Ω, and consequently $f^{\complement V}$ is finely l.s.c. and $\geqslant -p$ in Ω, finely hyperharmonic in V, and finely harmonic in $V \cap [f^{\complement V} < +\infty]$.

Clearly $u^{\complement V}$ is defined in U, and $u^{\complement V} = f^{\complement V}$ in $\tilde{V}$. In fact, $\varepsilon_x^{\complement V}$ is carried by $\partial_f V$ for $x \in V$ (§4.9), and also for $x \in \partial_f V \subset \complement V$ where $\varepsilon_x^{\complement V} = \varepsilon_x$. Thus $u^{\complement V}$ is finely l.s.c. relatively to $\tilde{V}$ and $\geqslant -p$ in $\tilde{V}$, finely hyperharmonic in V, and finely harmonic in $V \cap [u^{\complement V} < +\infty]$, all because $u^{\complement V} = f^{\complement V}$ in $\tilde{V}$. Since $u^{\complement V} = u$ in $U \setminus V$, we conclude that $u^{\complement V}$ is indeed finely hyperharmonic in the whole of U by the above lemma applied to the functions u (in U) and $v := u^{\complement V}$ (in V).

For any finely hyperharmonic function w in U such that $w \geq u$ in $U \setminus V$ and $w \geq -\mu$ in V, we similarly obtain from Lemma 9.5

$$w \geq w^{\complement V} \geq u^{\complement V} \qquad \text{in } V ,$$

and consequently $w \geq u^{\complement V}$ in all of U since $u^{\complement V} = u$ in $U \setminus V$. ▯

Corollary. For any open set U in the initial topology the finely harmonic (resp. finely hyperharmonic) functions in any finely open subset of U are the same whether considered relatively to the harmonic space Ω or the harmonic subspace U .

Recall at this point that U (with the restricted sheaf of harmonic functions) is again a strong harmonic space satisfying axiom (D), and that the fine topology and the notion of thinness in U are induced by the fine topology and the thinness in Ω. To verify the corollary it therefore suffices to prove that, for every regular (cf. §8.4) finely open set V with $\tilde{V} \subset U$ and for every $x \in V$, the swept-out measure $\varepsilon_x^{\complement V}$ relative to Ω is likewise the swept-out of ε_x on $U \setminus V$ relative to the harmonic space U . Explicitly this means that, for any hyperharmonic function $u \geq 0$ in U , the function $u^{\complement V}$ defined in U by $u^{\complement V}(x) = \int u \, d\varepsilon_x^{\complement V}$ should be the smallest hyperharmonic function ≥ 0 in U majorizing u in $U \setminus V$. And this is indeed the case according to the above theorem in view of Theorem 9.8.

10.3. <u>Lemma</u>. Let u be finely hyperharmonic in U (finely open), and suppose that u admits at least one finite valued, finely hypoharmonic minorant in U. The pointwise supremum (within $[-\infty, +\infty]^U$) of all finely hypoharmonic minorants for u in U is then finely harmonic in the finely open set $\{x \in U \mid u(x) < +\infty\}$.

Under the further hypothesis $u < +\infty$ in U, u therefore admits a greatest finely hypoharmonic minorant h in U, and h is finely harmonic in U.

<u>Proof</u>. Consider a fixed regular finely open set V of class $\mathcal{D}(u)$ relatively to U (see Def. 8.4). For any finite valued, finely hypoharmonic function $v \leq u$ (in U) it follows from Lemma 9.5 (since $V \in \mathcal{D}(v^+)$) that

$$v \leq v^{\complement V} \leq u^{\complement V} \leq u \qquad \text{in } V .$$

The inequality $v^{\complement V} \leq u$ holds in all of U because $v^{\complement V} = v$ in $U \setminus V$. According to the preceding theorem $v^{\complement V}$ is finely hypoharmonic (and $\geq v > -\infty$) in all of U, and finely harmonic in V.

When v ranges over the class, $\mathcal{M}$, of all finite valued, finely hypoharmonic minorants for u in U, we therefore have, for every $x \in U$,

$$h(x) := \sup_{v \in \mathcal{M}} v(x) = \sup_{v \in \mathcal{M}} v^{\complement V}(x) .$$

Clearly $-\infty < h \leq u < +\infty$ in V because $V \in \mathcal{D}(u)$. Invoking Theorem 9.11 we conclude that h is finely harmonic in V.

Since u is finite and finely continuous (hence finely locally bounded) in the finely open subset $[u < +\infty]$ of U, the sets V in question form a basis for the fine topology in $[u < +\infty]$ (§8.4), and consequently h is finely harmonic in $[u < +\infty]$ by the sheaf property (§8.6).

10.4. Definition. A finely hyperharmonic function u in U (finely open) is said to be finely superharmonic if $u < +\infty$ quasi everywhere in U, that is, if $[u = +\infty]$ is polar.

We denote by $\mathcal{S}(U)$ the set of all finely superharmonic functions $\geqslant 0$ in U.

According to Theorem 12.9 below a finely hyperharmonic function u (in U) is finely superharmonic if and only if u is not identically $+\infty$ in any fine component of U, or equivalently if u is finite in a finely dense subset of U.

In the case of an open set U in the initial topology the usual superharmonic functions in U are precisely those finely superharmonic functions in U which are locally bounded from below in U. This follows from Theorem 9.8. In particular, $\mathcal{S}(\Omega) = \mathcal{S}$, the usual superharmonic functions $\geqslant 0$ in U.

10.5. Definition. A fine potential relative to a finely open set $U \subset \Omega$ is a finely superharmonic function $p \geqslant 0$ in U such that every finely hypoharmonic minorant for p in U is $\leqslant 0$.

The latter requirement is equivalent to saying that 0 is the only finely hypoharmonic minorant $\geqslant 0$ for p. (In fact, the positive part of a finely hypoharmonic function is finely hypoharmonic.)

If $p \in \mathcal{S}(U)$ is finite valued, it follows from the second assertion of Lemma 10.3 that the words "finely hypoharmonic" may be replaced by "finely harmonic" in either formulation. The question remains open whether the same applies in general.

10.6. Theorem. The fine potentials relatively to U (finely open) form a face in the lattice cone of all finely hyperharmonic functions $\geqslant 0$ in U (under pointwise order), and so do the finely superharmonic functions $\geqslant 0$ in U.

Proof. The latter assertion is obvious. To prove (in the usual way) that the sum of two fine potentials p and q is a fine potential, let v denote a finely hypoharmonic minorant for $p + q$. Then $v - p$ is a finely hypoharmonic minorant for q, hence $v - p \leqslant 0$, and consequently $v \leqslant 0$. The remaining properties of a face are easily checked.

Remark. More generally, any sum p of (infinitely many) fine potentials relative to U is again a fine potential relative to U provided that p is at all finely superharmonic in U (that is, if $p < +\infty$ q.e. in U). The proof from the usual theory carries over, cf. [22 bis, Prop. 2.2.2].

10.7. Theorem. Every finite valued, finely hyperharmonic function $s \geqslant 0$ in U (finely open) has a unique decomposition of the form $s = p + h$ such that p is a (finite) fine potential relative to U and h is finely harmonic in U.

Proof. The existence of such a decomposition follows from the latter assertion of Lemma 10.3, according to which s admits a greatest finely hypoharmonic minorant h, and h is finely harmonic. Clearly $p := s - h$ is a fine potential relative to U.

The uniqueness is established in the usual simple way as follows: Let $s = p_1 + h_1$ be a further decomposition of the required type. Then $p + h = s \geq h_1$, that is, $h_1 - h \leq p$. Since $h_1 - h$ is a finely harmonic minorant for the fine potential p, we get $h_1 - h \leq 0$. Similarly $h - h_1 \leq 0$. ▯

__Remark__. The restriction to a __finite__ s is irrelevant for the uniqueness of such a decomposition, but cannot be weakened to the (necessary) requirement that s be finely superharmonic. This appears from the example in §8.10. In fact, let μ_n denote the trace of $\mu = \varepsilon_x^A$ on $\complement \omega_n$, where (ω_n) is a decreasing sequence of open neighbourhoods of x (in the initial topology on Ω) with the intersection $\{x\}$. Since $\mu(\{x\}) = 0$ we then have $G\mu = \sup G\mu_n$ with $G\mu_n$ finely harmonic in U by Theorem 8.10. The restriction s of $G\mu$ to U has therefore no greatest finely harmonic minorant in U because $s(x) = +\infty$.

10.8. __Theorem__. (The relative fine boundary minimum principle without exceptional points.) Let u be finely hyperharmonic in a finely open subset V of a finely open set U. Suppose that

$$\operatorname*{fine\ lim\ inf}_{x \to y,\ x \in V} u(x) \geq 0 \qquad \text{for every } y \in U \cap \partial_f V.$$

If moreover $u \geq -p$ in V for some __fine__ potential p relative to U, then $u \geq 0$ (in V).

__Proof__. In view of Lemma 10.1 the extension of u^- by 0 in $U \setminus V$ is a finely hypoharmonic minorant for p in U. ▯

10.9. Definition. A fine potential p relative to a finely open set U is said to be stable if, for every polar set e, the restriction of p to $U \setminus e$ is a fine potential (relative to $U \setminus e$).

Not every fine potential is stable (cf. Theorem 10.12).

Lemma. Each of the following conditions is necessary and sufficient for a finely superharmonic function $p \geqslant 0$ in U to be a stable, fine potential relative to U:

i) The restriction of p to $[p < +\infty]$ is a fine potential.

ii) 0 is the only minorant of class $\mathcal{S}(U)$ for p in U which is finely harmonic q.e. in U (that is, finely harmonic in $U \setminus e$ for some polar set e).

Proof. We shall repeatedly use Theorem 9.14 on removable singularities for finely hyperharmonic functions (in casu of class $\mathcal{S}(U)$). With this in mind it becomes clear that every stable, fine potential p satisfies ii), and also that ii) implies i). Now suppose that i) holds, and consider any polar set e and any finely hypoharmonic minorant v for p in the finely open set $U \setminus e$. Since v is finite and finely continuous in $[p < +\infty] \setminus e$, the restriction of v to this set extends by fine continuity to a finely hypoharmonic function $v' \leqslant p$ in $[p < +\infty]$. It follows from i) that $v' \leqslant 0$ in $[p < +\infty]$. Hence $v \leqslant 0$ in $[p < +\infty] \setminus e$, and so in $U \setminus e$ by fine continuity. ∎

Corollary. Every finite valued, fine potential is stable. Every sum of stable, fine potentials is a stable fine potential if at all finely superharmonic (that is, finite q.e.). (Cf. §10.6.)

In particular, every sum of finite, fine potentials is a stable, fine potential if it is finite q.e. Conversely, every stable, fine potential is a sum of finite ones, see Theorem 11.18.

10.10. Theorem. Every finely superharmonic function $s \geq 0$ in U (finely open) has a unique decomposition of the form $s = p + h$ such that p is a stable fine potential relative to U whereas h is of class $\mathcal{S}(U)$ and finely harmonic q.e. in U (hence finely harmonic in $[h < +\infty]$).

Proof. Let $s_1 = p_1 + h_1$ be the decomposition of the restriction s_1 of s to $[s < +\infty]$ into its fine potential part p_1 and its finely harmonic part h_1 according to Theorem 10.7. Denote by p and h the extensions of p_1 and h_1, respectively, by fine continuity to all of U (Theorem 9.14). Then p, $h \in \mathcal{S}(U)$, and $s = p + h$. Clearly p and h have the stated properties.

Conversely, let $s = p + h$ denote an arbitrary decomposition of the kind in question. Using an index 1 to designate restriction to $[s < +\infty]$, p_1 is a fine potential relative to $[s < +\infty]$, and h_1 is finely harmonic by Theorem 9.15 or §9.16. Consequently p_1 and h_1 are the fine potential part and the finely harmonic part of s_1, respectively, and so p and h are uniquely determined as the extensions of p_1 and h_1 by fine continuity. ▯

10.11. Theorem. (The relative fine boundary minimum principle with exceptional points.) Let u be finely hyperharmonic in a finely open subset V of a finely open set U. Suppose that

$$\operatorname{fine}\liminf_{x \to y,\, x \in V} u(x) \geq 0 \qquad \text{for quasi every } y \in U \cap \partial_f V.$$

If moreover $u \geq -p$ in V for some stable fine potential p relative to U, then $u \geq 0$ (in V).

Proof. Let e denote the polar set of those points $y \in U \cap \partial_f V$ at which the above boundary inequality fails to hold, and apply Theorem 10.8 to u (in V) and to the restriction p_1 of p to the finely open set $U \setminus e$ ($\supset V$), noting that p_1 is a fine potential relative to $U \setminus e$. ▯

Remark. For any given finely open set U the above theorem would no longer hold if p were replaced by any function s in $\mathcal{S}(U)$ which is not a stable, fine potential. In fact, let $s = p + h$ be the decomposition of s from the preceding theorem, and take $u = -h$, $V = [h < +\infty]$ in the assertion of the present theorem (with s in place of p), observing that $U \cap \partial_f V = [h = +\infty]$ is polar.

Theorem 10.11 contains the fine boundary minimum principle given in Theorem 9.1 corresponding to the case $U = \Omega$. This follows from Theorem 10.12 below. Consequently, the requirement in Theorem 9.1 that p be a semibounded potential cannot be relaxed within the class of superharmonic functions $\geqslant 0$ in Ω .

Corollary. Every stable, fine potential p relative to U (finely open) has the following fine domination property:

Let p be finely harmonic q.e. in $U \setminus S$ for some $S \subset U$, S finely closed relative to U , and let $u \geqslant 0$ be finely hyperharmonic in U . If the inequality $u \geqslant p$ holds q.e. in S , then it holds everywhere in U .

Adding, if necessary, to S the polar set $[p = +\infty]$, we may assume that p is finely harmonic in all of $V := U \setminus S$. Now apply Theorem 10.11 to the finely hyperharmonic function $u - p$ ($\geqslant -p$) in V , observing that, by the fine continuity of u and p , the boundary inequality in the theorem is satisfied q.e. in $U \cap \partial_f V$, viz. at any point $y \in U \cap \partial_f V$ $(\subset S)$ such that $u(y) \geq p(y)$ and $p(y) < +\infty$.

10.12. In the rest of the present section we shall specialize to the case of an open set U in the <u>initial</u> topology. In view of Cor. 10.2 it suffices to consider the case $U = \Omega$. We shall compare the above notion of a fine potential relative to Ω with the usual notion of a potential on Ω.

<u>Theorem</u>. The notion of a <u>stable</u>, fine potential relative to Ω is identical with that of a <u>semibounded</u> (ordinary) potential on Ω.

<u>Proof</u>. In the first place, <u>any</u> fine potential relative to Ω is an ordinary potential (see the next theorem). Now consider a stable, fine potential p relative to Ω. To show that p is semibounded, consider the swept-out potentials

$$\hat{R}_p^{[p \geq \lambda]} = p^{[p \geq \lambda]} \qquad (\lambda \text{ real}).$$

Each of these is finely harmonic in the finely open set $[p < \lambda]$ according to Cor. 9.7 (or Lemma 9.12, or Theorem 9.13). Hence

$$h := \inf_\lambda \hat{R}_p^{[p \geq \lambda]}$$

is finely harmonic in $[p < +\infty]$ by virtue of Theorem 9.11. Since $0 \leq h \leq p$, and p is a fine potential relative to $[p < +\infty]$, we conclude that $h = 0$ in $[p < +\infty]$. The l.s.c. envelope $\hat{h}$ of h is therefore $= 0$ in $[p < +\infty]$, and hence in all of Ω by fine continuity. Consequently p is semibounded (Def. 2.1).

Conversely, every semibounded potential s on Ω is of class $\mathcal{S} = \mathcal{S}(\Omega)$, and s must be a stable, fine potential relative to Ω in view of the original fine boundary minimum principle (Theorem 9.1) combined with the remark to Theorem 10.11.

Corollary. The notion of a finite valued, fine potential relative to Ω is identical with that of a finite and semibounded potential on Ω .

10.13. Theorem. Every fine potential relative to Ω is an ordinary potential on Ω . The converse holds e.g. if Ω is a Green space.

Proof. If p is a fine potential relative to Ω , then $p \in \mathcal{S}(\Omega) = \mathcal{S}$ (§10.4), that is, p is an ordinary superharmonic function $\geqslant 0$ in Ω . Any subharmonic minorant for p in Ω is finely hypoharmonic and hence $\leqslant 0$.

Conversely, let p denote an ordinary potential on Ω . Then p is finely superharmonic and $\geqslant 0$. In view of Theorem 10.10 (together with Theorem 10.6) it suffices to consider the case where p is finely harmonic in $[p < +\infty]$. In this case, let v denote any finely hypoharmonic minorant $\geqslant 0$ for p in Ω . Then

$$q := p - v$$

is finely superharmonic and $\geqslant 0$ in Ω , hence of class $\mathcal{S}$. Since $q \leqslant p$, q is even an ordinary potential. For fixed $k > 1$ the set

$$A := [p > kv]$$

is finely open because v is finely continuous. Clearly

$$A \subset b(A) = \tilde{A} \subset [p \geq kv],$$

$$q \geq \frac{k-1}{k} p \quad \text{in } A,$$

$$q \geq \frac{k-1}{k} \hat{R}_p^A \quad \text{in } \Omega.$$

The proof will be complete if it can be shown that $\hat{R}_p^A = p$ for every choice of $k > 1$, as this will imply $q = p$, $v = 0$.

For this purpose, choose a potential $r > 0$ of class $\mathcal{P}^c$ on Ω, and consider the sequence of finely open sets

$$V_n := [p < nr].$$

Clearly $\tilde{V}_n \subset [p < +\infty]$. Since nr is a semibounded potential, it follows from Lemma 9.5 (applied to the finely harmonic functions p and $-p$ in the finely open set $[p < +\infty]$) that $p = p^{\complement V_n}$ in V_n. The same equality holds trivially in $b(\complement V_n)$, thus altogether everywhere in Ω by fine continuity.

Now, if Ω is a Green space with Green kernel G, we may write $p = G\lambda$, $q = G\mu$ with admissible measures λ, μ on Ω. The relation $G\lambda = (G\lambda)^{\complement V_n} = G(\lambda^{\complement V_n})$ shows that $\lambda = \lambda^{\complement V_n}$, that is, λ is carried by $b(\complement V_n) \subset \complement V_n$, and so $\lambda(V_n) = 0$. Since the sequence (V_n) covers $[p < +\infty]$, it follows that λ is carried by $[p = +\infty]$. In particular, λ is carried by the above finely open set $A = [p > kv]$ and hence by $b(A)$. We conclude that $\lambda^A = \lambda$ (see §4.7), and so

$$\hat{R}_p^A = (G\lambda)^A = G(\lambda^A) = G\lambda = p. \blacksquare$$

10.14. Lemma. The following 4 propositions imply each other (for any strong harmonic space satisfying axiom (D)). They hold true in the case of a Green space.

a) Every potential on Ω is a fine potential relative to Ω.

b) Every potential on Ω which is finely harmonic quasi everywhere in Ω, is a fine potential relative to Ω.

c) The fine boundary minimum principle given in Theorem 9.1 remains valid with p replaced by an arbitrary potential on Ω, provided that the boundary inequality is required to hold everywhere on $\partial_f U$.

d) For every potential p on Ω the relation $R_p^S = p$ holds for every finely closed set S such that p is finely harmonic in $\complement S$ and finite on $\partial_f S$.

Proof. It was observed in the beginning of the proof of Theorem 10.13 that a) $\Longleftrightarrow$ b). From Theorem 10.8 follows a) $\Longrightarrow$ c). The implication c) $\Longrightarrow$ d) is established in much the same way as Cor. 10.11, as follows: Let $u \geq 0$ be hyperharmonic in Ω, and suppose that $u \geq p$ in S. Then $u - p$ is finely hyperharmonic and $\geq -p$ in $U := \complement S$, and finely continuous in $\tilde{U}$ (because u and p are finely continuous, and p is finite in $\tilde{U}$). Hence it follows from c) that $u - p \geq 0$ in $U = \complement S$, and so $u \geq p$ in all of Ω, showing that indeed $R_p^S = p$.

Finally it appears from the proof of Theorem 10.13 that d) $\Longrightarrow$ b). In fact, let the potential p on Ω be finely harmonic in $[p < +\infty]$. The finely open set A in the proof in question contains the finely closed set $S := [p \geq kv + 1]$ because v is finite. Since p is finite on $\partial_f S$ and finely harmonic in $\complement S$, it follows from d) that $\hat{R}_p^A \geqslant \hat{R}_p^S = R_p^S = p$.

In the greenian case, a) holds by the preceding theorem. ▯

Remark. The four equivalent propositions in the above lemma hold good, not only for a Green space, but also for a harmonic space in Brelot's case A_2 (see §2.7). The proof given in Theorem 10.13 carries over. For a still more general case see §10.17 below.

Theorems 10.15 and 10.16 below therefore likewise hold in case A_2. The same applies to Theorem 8.10, and altogether to all results stated in the present memoir for a Green space.

10.15. Theorem. The following 4 propositions imply each other (for any strong harmonic space Ω satisfying axiom (D)). They hold true in the case of a Green space.

a) Every finite potential p on Ω is semibounded (or, equivalently: is a fine potential relative to Ω).

b) Every potential on Ω which is finely harmonic in Ω is identically 0 .

c) Every finely harmonic function on Ω having a harmonic minorant (or majorant) is itself harmonic in Ω .

d) A finite potential p on Ω (say of compact harmonic support $S(p)$) is continuous if p is continuous relative to $S(p)$.

Proof. As to the alternative formulation of a), see Theorem 10.12 (and Cor. 10.9). The implication a) $\Longrightarrow$ b) is obvious since a finely harmonic function is finite valued.

As regards b) $\Longrightarrow$ c), let u be finely harmonic in Ω , and let $u \geqslant v$, where v is harmonic (hence also finely harmonic) in Ω . Then $u - v$ ($\geqslant 0$) is finely harmonic in Ω , hence hyperharmonic there, by Theorem 9.8, and indeed superharmonic, being finite valued. The usual harmonic part, h , of $u - v$ is also finely harmonic, and so is therefore the usual potential part $p := u - v - h$ of $u - v$. Hence it follows from b) that $p = 0$, whence $u = v + h$ is harmonic in Ω .

To show that c) $\Longrightarrow$ a), consider a finite potential p on Ω . Then p (≥ 0) is finite and finely superharmonic (§10.4). Let h denote the finely harmonic part of p in Ω (Theorem 10.7). Under the hypothesis c), h is harmonic and consequently $= 0$. This shows that p is a fine potential (and a stable one by Cor. 10.9).

To show that a) $\Longrightarrow$ d), let $q \in \mathcal{P}^b$, $q > 0$. A finite potential p with a continuous restriction to its compact harmonic support $S(p)$ is majorized on $S(p)$ by kq for some constant k. If p is semibounded, p has the domination property (Theorem 9.2), and hence $p \leq kq$ in all of Ω . Consequently p is continuous according to the continuity principle for locally bounded superharmonic functions [8, Theorem 26]. The assumption that $S(p)$ be compact can be removed by application of the partition theorem [29, Théorème 12.2] to a compact neighbourhood of any given point of Ω .

Finally the implication d) $\Longrightarrow$ a) follows from the fact that every finite potential p is representable as the specific supremum of those specific minorants q of p of compact harmonic support $S(q)$ for which q is continuous relatively to $S(q)$. See [5, Théorème 1], and also §2.5 above.

<u>Remark</u>. Not every strong harmonic space Ω with axiom (D) is such that the four equivalent propositions in the above theorem

hold true.[23] A fortiori, the four equivalent properties listed in Lemma 10.14 are not shared by every such space. This appears from the example given in Constantinescu and Cornea [22, Cor. 1.2]. In this example a finite potential, t_z , of one point harmonic support $\{x_0\}$ is so constructed that t_z is discontinuous at x_0 .

It is well known that, for any strong harmonic space Ω , the proposition d) in Theorem 10.15 (the continuity principle for finite potentials) is equivalent to saying that every finite potential has the domination property (§1.10). Hence this proposition implies the domination axiom (D) (but not conversely).

10.16. Theorem. In a Green space Ω , a finely harmonic function u in an open subset U of Ω is harmonic if (and only if) u is locally bounded from one side in U .

Proof. The proof merely depends on the validity of the proposition c) in the preceding theorem for every open subspace of Ω . Suppose that u is finely harmonic in the open set U and locally bounded from below there. Every point of U has then an open neighbourhood V in which u is minorized by a harmonic function v . Consequently u is harmonic in V by virtue of c) of Theorem 10.15, applied to the harmonic subspace V (cf. at this point Cor. 10.2). - For the converse, see Theorems 8.7 or 9.8. ▯

---

23) In [26, note 4, p. 16] and in [28, p. 514] the present author has incorrectly stated that every finite potential is semibounded in the general case of a strong harmonic space with axiom (D). Accordingly, the term "finite potential" should be replaced by "finite and semibounded potential" several times in [28](unless the attention is restricted e.g. to the greenian case). Similarly as to the form of Theorem 10.16 above in [28

Remark. Actually, Theorem 10.16 remains valid with the Green space replaced by any strong harmonic space in which the continuity principle for finite potentials holds, in other words (cf. Remark 10.15) any strong harmonic space with axiom (D) in which the four equivalent propositions in Theorem 10.15 hold true. Thus the property stated in Theorem 10.16 may be added to the list of equivalent propositions in Theorem 10.15. This is because property a) in Theorem 10.15 is preserved by passage to open subspaces of Ω , as it can be shown by application of the extension theorem of Mme Hervé.

Problem. Can the one sided local boundedness condition in Theorem 10.16 be dropped (in the greenian case). In other words: is every finely harmonic function on a Green space harmonic? More generally: is every finely hyperharmonic function on a Green space hyperharmonic? - I am inclined to believe that the answers are no.

10.17. It follows from Taylor [36, Theorem 5.4] that every strong harmonic space satisfying the domination axiom (D) and the axiom of proportionality (P) (see a), §2.7 above) admits a dual process in the sense of Kunita and Watanabe. Hence it follows from Nguyen-Xuan-Loc [32, Theorem 10] (see also [34]) that every strong harmonic space with (D) and (P) has property c) in Lemma 10.14 above. Thus all potentials on such a space are fine potentials.

Furthermore it is easily shown that every strong harmonic space with (D) having the equivalent property d) in Lemma 10.14 satisfies the continuity principle for general potentials (not only for finite potentials as in d) of Theorem 10.15).

11. Balayage and specific multiplication relative to a finely open set

We continue the study of concepts relative to a fixed, finely open set $U \subset \Omega$. For any numerical function f on U we shall allow ourselves in the present section to let $\hat{f}$ denote the finely l.s.c. envelope of f, defined in U by

$$\hat{f}(x) = \underset{y \to x,\, y \in U}{\text{fine lim inf}}\ f(y), \qquad x \in U.$$

11.1. Definition. A numerical function v defined in U is called nearly finely hyperharmonic if v is finely locally bounded from below in U, and if

$$v(x) \geq \int^{*} v \, d\varepsilon_x^{\complement V} \qquad \text{for every} \quad x \in V \in \mathcal{D}(v^-).$$

As in §8.4, $\mathcal{D}(v^-)$ denotes the class of all finely open sets V of compact closure $\overline{V} \subset U$ such that v^- is bounded on $\overline{V}$. Since v^- is finely locally bounded in U, $\mathcal{D}(v^-)$ is a basis for the fine topology in U.

Clearly it suffices to verify the condition in the above definition for regular sets V of class $\mathcal{D}(v^-)$.

Every finely hyperharmonic function in U is also nearly finely hyperharmonic according to Lemma 9.5.

The pointwise infimum (within $[-\infty, +\infty]^{\Omega}$) of any family of nearly finely hyperharmonic functions is nearly finely hyperharmonic if it is finely locally bounded from below.

11.2. Lemma. The finely l.s.c. envelope $\hat{v}$ of a nearly finely hyperharmonic function v in U is finely hyperharmonic in U. For every $x \in U$

$$\hat{v}(x) = \sup \{ \int^* v \, d\varepsilon_x^{\complement V} \mid V \in \mathcal{D}(v^-), \; V \ni x \}.$$

Proof. The usual proof carries over (see e.g. Bauer [1, Satz 2.1.1] where $\mathcal{D}$ is to be replaced now by $\mathcal{D}(v^-)$). In fact, $\int^* v \, d\varepsilon_x^{\complement V}$ is finely hyperharmonic as a function of $x \in V$ by Theorem 9.13, and every usual harmonic function h in a usual open set $\omega \subset \Omega$ is finely harmonic in ω, whence $\int h \, d\varepsilon_x^{\complement V} = h(x)$ for every $V \in \mathcal{D}(v^-)$ with $x \in V \subset \overline{V} \subset \omega$ (Lemma 9.5). ▮

11.3. Lemma. Let (v_n) denote a sequence of nearly finely hyperharmonic functions ≥ 0 in Ω.

a) $\sum_n v_n$ is nearly finely hyperharmonic, and

$$\widehat{\sum_n v_n} = \sum_n \hat{v}_n .$$

b) If (v_n) is pointwise increasing, then the pointwise supremum, $\sup_n v_n$, is nearly finely hyperharmonic, and

$$\widehat{\sup_n v_n} = \sup_n \hat{v}_n .$$

Proof. Ad b). Clearly $v := \sup v_n$ is nearly finely hyperharmonic. The stated formula for $\hat{v}$ follows from the preceding lemma by interchanging the two suprema.

Ad a). In view of b) it suffices to consider the case of a finite sequence, or just 2 nearly finely hyperharmonic functions v_1, v_2. Clearly $v_1 + v_2$ is nearly finely hyperharmonic. The stated formula for $\widehat{v_1 + v_2}$ follows again by use of the above lemma, the inequality $(v_1 + v_2)^\wedge \geqslant \hat{v}_1 + \hat{v}_2$ being obvious. ▯

<u>Corollary</u>. For any sequence of finely hyperharmonic (or just nearly finely hyperharmonic) functions $u_n \geqslant 0$ in U the function

$$u \ := \ \liminf_n \ u_n$$

is nearly finely hyperharmonic in U, and its finely l.s.c. envelope is

$$\hat{u} \ = \ \sup_n \widehat{\inf_{p>n} u_p} \ .$$

11.4. We denote by $\mathcal{U}(U)$ the lattice cone of all finely hyperharmonic functions $\geqslant 0$ in the given finely open set U.

<u>Definition</u>. Let $A \subset U$, and let $f \geqslant 0$ be defined at least in A. The corresponding reduced function relative to U, denoted by ${}_U R_f^A$, or simply by R_f^A, is defined in U by

$${}_U R_f^A(x) \ = \ R_f^A(x) \ = \ \inf\{\, u(x) \mid u \in \mathcal{U}(U),\ u \geqslant f \text{ in } A\}.$$

The swept-out function relative to U, denoted by ${}_U \hat{R}_f^A$, or simply by $\hat{R}_f^A$, is defined in U as the finely l.s.c. envelope of ${}_U R_f^A$:

$${}_U \hat{R}_f^A(x) \ = \ \hat{R}_f^A(x) \ = \ \text{fine} \liminf_{y \to x,\, y \in U} \ {}_U R_f^A(y) \ .$$

If $A = U$ we may write simply ${}_U R_f$ and ${}_U \hat{R}_f$, respectively, or even just R_f and $\hat{R}_f$, in place of ${}_U R_f^U$ and ${}_U \hat{R}_f^U$, respectively.

Note that these definitions agree with the usual ones in the case $U = \Omega$ on account of Theorem 9.8, according to which $\mathcal{U}(\Omega) = \mathcal{U}$ (the usual hyperharmonic functions ≥ 0), together with the fact that the usual swept-out function $\hat{R}_f^A$ is the l.s.c. envelope of R_f also in the fine topology (see e.g. [1, Lemma 3.1.5]).

In the rest of this section R_f^A and $\hat{R}_f^A$ (and similarly R_f and $\hat{R}_f$) always refer to these notions taken relatively to the given finely open set $U \subset \Omega$. The corresponding concepts relative to the whole space Ω will therefore be denoted by ${}_{\Omega}R_f^A$, ${}_{\Omega}\hat{R}_f^A$, etc. (instead of the usual notation R_f^A, $\hat{R}_f^A$).

Just like in the usual situation, R_f^A is (pointwise) increasing and countably subadditive in each argument A or f. The same applies to $\hat{R}_f^A$ in view of Lemma 11.3.

Note also that $R_u^A \leq u$ (in U), with equality in A , for any $u \in \mathcal{U}(U)$ and any $A \subset U$.

The following results from §3.7 extend immediately to the present more general case:

Let f denote a numerical function ≥ 0 in U , and write $\tilde{f}$ for the finely u.s.c. envelope of f in U . Then

$$R_{\tilde{f}} = R_f \quad ; \quad \hat{R}_{\tilde{f}} = \hat{R}_f .$$

If $f \geq 0$ is finely l.s.c. then

$$\hat{R}_f = R_f ,$$

and for any set $A \subset U$

$$R_f^{\tilde{A} \cap U} = R_f^A \quad ; \quad \hat{R}_f^{\tilde{A} \cap U} = \hat{R}_f^A .$$

11.5. Lemma. Let B denote a base in Ω contained in U. Let f be a finely continuous function defined and $\geqslant 0$ in B, and majorized there by some finite and semibounded potential p on Ω. Then $\hat{R}_f^B = R_f^B$.

Proof. We extend f by 0 in $b(\complement U)$ $(\subset \complement U)$. Clearly f remains finely continuous with $0 \leqslant f \leqslant p$ in the extended set of definition

$$B \cup b(\complement U) = b(B \cup \complement U).$$

According to Lemma 9.3 the function

$$v := f^{B \cup b(\complement U)}$$

is a finely continuous further extension of f to all of Ω, and $0 \leqslant v \leqslant p$ there. Moreover, v is finely harmonic off $B \cup b(\complement U)$, thus in particular in $U \setminus B$.

To calculate R_f^B, consider any $u \in \mathcal{U}(U)$ such that $u \geqslant f$ in B (or equivalently $u \geqslant v$ in B). Then $u - v$ is finely hyperharmonic in $U \setminus B$. Since $u \geqslant 0$, and since v is finely continuous in Ω, we obtain

$$\underset{x \to y,\ x \in U \setminus B}{\text{fine lim inf}} \big(u(x) - v(x)\big) \geqslant -v(y)$$

for every $y \in \partial_f(U \setminus B) = (\partial_f U) \cup (\partial_f B)$. For every $y \in \partial_f B$ this fine lim inf equals $u(y) - v(y) \geqslant 0$ since $B \subset U$. For every regular fine boundary point $y \in (\partial_f U) \cap b(\complement U)$ for U we have $v(y) = f(y) = 0$, and hence the above fine lim inf is $\geqslant 0$ in this case, too, thus altogether q.e. on $\partial_f(U \setminus B)$. It follows now from the fine boundary minimum principle (Theorem 9.1) that $u - v \geqslant 0$, that is, $u \geqslant v$, in $U \setminus B$, and hence in all of U . Consequently,

$$R_f^B = R_v \quad ; \quad \hat{R}_f^B = \hat{R}_v .$$

However, $R_v = \hat{R}_v$, as noted in §11.4, because v is finely continuous.

11.6. Lemma. For any usual polar set $e \subset U$, we have $R_{+\infty}^e = 0$ in $U \setminus e$, and $\hat{R}_{+\infty}^e = 0$ in U.

Proof. For any given $x \in U \setminus e$ there exists a usual superharmonic function $s \geq 0$ on Ω such that $s = +\infty$ in e , whereas $s(x) < +\infty$. Since s is likewise finely superharmonic in Ω , in particular in U , the former assertion is obtained by considering the functions εs $(\geq +\infty \cdot 1_e)$ for small ε . The latter assertion follows from the former because e has no finely interior points.

Corollary. For any function $f \geq 0$ the swept-out function $\hat{R}_f$ relative to U does not change if f is changed in a polar set. In particular,

$$\hat{R}_f^A = \hat{R}_f^{\tilde{A} \cap U} = \hat{R}_f^{b(A) \cap U}$$

for any set $A \subset U$ and any finely l.s.c. function $f \geq 0$ in U .

11.7. It is a simple consequence of the fine continuity of the functions of class $\mathcal{U}(U)$ that a set $A \subset U$ is thin at a point $x \in U \setminus A$ if and only if A is "thin at x relatively to U " in the sense that there exists a fine neighbourhood V of x in U such that

$$R_1^{A \cap V}(x) < 1 .$$

This follows, in fact, from a general result of Brelot [10, Théorème 5].

The following parallel result concerning the characterization of the thinness of a set $A \subset U$ (even at points of A) as "weak thinness relative to U " is, however, less obvious.

<u>Theorem</u>. In order that a subset A of U be thin at a point $x \in U$ it is necessary and sufficient that there exists a fine neighbourhood V of x in U such that $\hat{R}_1^{A \cap V}(x) < 1$.

<u>Proof</u>. Recall the convention $\hat{R}_f := {}_U\hat{R}_f$, etc., in the present situation (§11.4). Suppose first that A is thin at x . Then x has a neighbourhood ω in the initial topology on Ω such that

$$ {}_{\Omega}\hat{R}_1^{A \cap \omega}(x) < 1. $$

It follows that $V := U \cap \omega$ is a fine neighbourhood of x in U , and that

$$ \hat{R}_1^{A \cap V}(x) \leq {}_{\Omega}\hat{R}_1^{A \cap V}(x) = {}_{\Omega}\hat{R}_1^{A \cap \omega}(x) < 1. $$

Conversely, as to the sufficiency, let ω denote any relatively compact, open neighbourhood of x in the initial topology on Ω. Then there exists a potential $p \in \mathcal{P}^c$ on Ω such that $p \geq 1$ on ω . Let V be as stated in the theorem. Replacing, if necessary, V by a suitable smaller set, we may assume, in addition, that $\tilde{V} \subset U \cap \omega$, the fine topology being regular.

Now write $B := b(A \cap V)$. Clearly

$$ B \subset b(V) = \tilde{V} \subset U \cap \omega , $$

and so $1 \leq p$ on B . We may therefore apply Lemma 11.5 to B and the function $f = 1$. In view of Cor. 11.6 this leads to

$$\hat{R}_1^{A \cap V} = \hat{R}_1^B = R_1^B ,$$

and consequently, by hypothesis, $R_1^B(x) < 1$, showing that $x \in \complement B$. This means, however, that $A \cap V$ is thin at x , and so is therefore A itself. ▯

11.8. <u>Theorem</u>. (The convergence theorem for finely hyperharmonic functions.) For any numerical function $f \geqslant 0$ on U ,

$$\hat{R}_f = R_f \qquad \text{quasi everywhere,}$$

that is, everywhere in U except possibly in some polar set.

Slightly more generally: The pointwise infimum $u := \inf_i u_i$ (within $[-\infty , +\infty]^U$) of a family of finely hyperharmonic functions u_i in U differs at most in a polar set from its finely l.s.c. envelope $\hat{u}$ in U , provided that u is finely locally bounded from below in U .

<u>Proof</u>. The exceptional set $[\hat{R}_f < R_f]$ is known to be a countable union of sets each of which is "weakly thin" with respect to the cone $\mathcal{U}(U)$ at each of its points (Brelot [12, Théorème I], Bauer [1, Satz 3.3.4]), or equivalently: thin in the usual sense, by the preceding theorem, and hence polar.

The latter version of the theorem is reduced to the former by replacing u_i (in a suitable fine neighbourhood V of a given point of U) by $u_i - h$ for some finely harmonic function $h \leqslant \inf_i u_i$ in V . ▯

<u>Corollary</u>. For any numerical function $f \geqslant 0$ on U the swept-out function $\hat{R}_f$ relative to U is the smallest function of class $\mathcal{U}(U)$ majorizing f quasi everywhere in U .

11.9. Theorem. The pointwise limit of a pointwise convergent sequence of finely harmonic functions u_n in U is finely harmonic provided that $\sup_n |u_n|$ is finely locally bounded in U.

Proof. The sign of inequality in the relations

$$u := \lim_n u_n = \sup_p \left(\inf_{n>p} u_n\right)$$
$$\geqslant v := \sup_p \left(\widehat{\inf_{n>p} u_n}\right)$$

may be replaced by equality quasi everywhere on account of the above convergence theorem. Note that v is finely hyperharmonic. Replacing u_n by $-u_n$, we similarly obtain $-u \geqslant w$ with equality q.e., w being finely hyperharmonic. Altogether, $v \leqslant u \leqslant -w$ with equality q.e., hence everywhere by the fine continuity of v and w. Thus $u = v$ and $-u = w$ are both finely hyperharmonic. ▯

11.10. Lemma. For any finely hyperharmonic function $u \geqslant 0$ in U, and for any set $A \subset U$,

$$\hat{R}_u^A = R_u^{b(A)\cap U} \qquad (= u \quad \text{on} \quad b(A)\cap U).$$

$\hat{R}_u^A$ is the smallest function of class $\mathcal{U}(U)$ which equals (or majorizes) u on the relative base $b(A)\cap U$ of A.

Proof. By Cor. 11.8 the finely open set $[\hat{R}_u^A < u]$ meets A in a polar set, hence does not meet $b(A)$. ▯

Remark. More generally, $\hat{R}_f = \hat{R}_{b(f)} = R_{b(f)}$ $(\geqslant b(f))$ for every numerical function $f \geqslant 0$ on U. Here $b(f)$ denotes the base function of Doob [23] relative to the fine topology in U, see also [25, §4.2] where also non polar points are admitted.

11.11. Lemma. For any numerical function $f \geq 0$ on U,

$$R_f(x) = \max\{f(x), \hat{R}_f(x)\} \quad \text{for every } x \in U.$$

Proof. The proof given in Lemma 4.5 for the particular case $U = \Omega$ carries over in view of Theorem 11.8 and Lemma 11.6. ▯

11.12. Theorem. For any increasing sequence of numerical functions $f_n \geq 0$ on U,

$$R_{\sup f_n} = \sup_n R_{f_n} \quad ; \quad \hat{R}_{\sup f_n} = \sup \hat{R}_{f_n} .$$

Proof. The proof given in Theorem 4.6 for the particular case $U = \Omega$ carries over in view of Theorem 11.8 and the above lemma. ▯

11.13. Theorem. Let $f \geq 0$ be defined in U and finely hypoharmonic in a finely open set $V \subset U$. Then $u := \hat{R}_f$ is finely harmonic in $V \cap [u < +\infty]$. Moreover, $u^{\complement W} = u$ for every finely open set W with $\tilde{W} \subset V$ such that $f \leq p$ in W for some semibounded potential p on Ω.

Proof. The proof given in Lemma 9.7 for the particular case $U = \Omega$ carries over in view of Theorem 10.2, according to which $u^{\complement W} = u^{b(\complement W)}$ is finely hyperharmonic in U. ▯

Corollary. For any finely superharmonic function $s \geq 0$ in U (Def. 10.4) and any set $A \subset U$, $\hat{R}_s^A$ is finely harmonic quasi everywhere in $U \setminus b(A)$ (hence everywhere in $\{x \in U \setminus b(A) \mid \hat{R}_s^A(x) < +\infty\}$).

11.14. It is now clear how one may continue extending results from the usual case $U = \Omega$ (or U open in the initial topology) to the present, more general case of reduction and balayage relative to a finely open set U, that is, taken with respect to the lattice cone $\mathcal{U}(U)$ of all finely hyperharmonic functions $u \geqslant 0$ in U.

Rather than carrying over the proofs from the usual theory, we shall avail ourselves of the general theory of balayage with respect to a suitable cone of continuous functions on a Baire space U as developed recently by Constantinescu and Cornea [22 bis, Chapter IV], see also Mokobodzki [31].

In the first place, Ω is known to be a Baire space also in the fine topology [22 bis, Cor. 5.1.1], and so is therefore the given finely open set U.

Let $\mathcal{W}$ denote the convex cone of all finely hyperharmonic (hence finely continuous by Theorem 9.10) functions on U. Thus $\mathcal{W}^+ = \mathcal{U}(U)$. According to Cor. 9.5, $\mathcal{W}$ is inf stable, and hence lower directed with respect to the pointwise order.

For any family of functions $u_i \in \mathcal{W}$, minorized in $\mathcal{W}$, the pointwise infimum $\inf u_i$ (relative to $\mathcal{F}(U) := [-\infty, +\infty]^U$) is nearly finely hyperharmonic in U (§11.1), and its finely l.s.c. envelope $\widehat{\inf u_i}$ is finely hyperharmonic in U (Lemma 11.2).

Thus the pointwise infimum, relative to $\mathcal{W}$, of any minorized family (u_i) on $\mathcal{W}$ exists and is given by

$$\bigwedge_i u_i = \widehat{\inf_i u_i} .$$

Since the constant $+\infty$ is the greatest element of $\mathcal{W}$, it follows that $\mathcal{W}$ is upper complete, and hence is a conditionally complete lattice with respect to the pointwise order. Hence $\mathcal{W}^+ = \mathcal{U}(U)$ is a complete lattice in the pointwise order.

Next, for any pointwise upper directed family (u_i) on $\mathcal{W}$, the pointwise supremum, $\sup_i u_i$, relative to $\mathcal{F}(U)$ belongs to $\mathcal{W}$ (Cor. 9.5), and hence equals $\bigvee_i u_i$, the pointwise supremum relative to $\mathcal{W}$. [24)]

Finally, $\mathcal{W}^+ = \mathcal{U}(U)$ has the <u>Riesz decomposition property</u> with respect to the pointwise order:

If $u \leqslant v + w$, there exist $v_1 \leqslant v$, $w_1 \leqslant w$ such that $u = v_1 + w_1$.

According to the following lemma, one may, in fact, take $w_1 = R_f$, where $f := (u - v)^+$; see [22 bis, Prop. 4.1.5].

Here and in the sequel we shall denote by $\preccurlyeq$ the <u>specific order</u> on $\mathcal{W}$, whereby $u_1 \preccurlyeq u_2$ means the existence of some $u \in \mathcal{U}(U)$ such that $u_1 + u = u_2$.

<u>Lemma</u>. Let $u, v \in \mathcal{U}(U)$, and put $f := (u - v)^+$. Then $R_f = \hat{R}_f \preccurlyeq u$.

<u>Proof</u>. The proof given in [22 bis, Theorem 5.1.1] for the case of usual hyperharmonic functions in Ω carries over. For the sake of completeness we bring the details.

---

24) As in the usual case, any minorized family (u_i) on $\mathcal{W}$ contains a countable subfamily with the same pointwise infimum $\bigwedge u_i$ relative to $\mathcal{W}$. This follows easily from the quasi Lindelöf principle (§3.9) in view of Theorem 11.8. - Also, any pointwise upper directed family (u_i) on $\mathcal{W}$ contains a countable subfamily with the same pointwise supremum $\bigvee u_i = \sup u_i$.

Choose an ordinary locally bounded potential $q > 0$ on Ω, and define a sequence of functions $u_n \in \mathcal{U}(U)$ by

$$u_n(x) = \min\{u(x),\ nq(x)\}, \qquad x \in U.$$

Each of the functions $(u_n - v)^+$ is finely continuous, and $f = (u - v)^+ = \sup_n (u_n - v)^+$ is therefore finely l.s.c. in U. According to §11.4 this implies that $\hat{R}_f = R_f$.

Now consider a fixed regular finely open set V with $\tilde{V} \subset U$. The function w defined in U by $w = u$ in $U \setminus V$, and

$$w = \inf\left\{ R_f,\ (R_f)^{\complement V} + u - u_n^{\complement V} \right\} \quad \text{in } V,$$

is finely hyperharmonic in U by Lemma 10.1 because $u_n^{\complement V}$ is finely harmonic in V, finely continuous in $\tilde{V}$, and equal to u_n $(\leq u)$ in $U \setminus V$ (see e.g. Lemma 9.3 or Theorem 10.2). Since $u_n \leq u \leq v + R_f$, we obtain in V, by Lemma 9.5,

$$u_n^{\complement V} \leq u^{\complement V} \leq v^{\complement V} + (R_f)^{\complement V} \leq v + (R_f)^{\complement V},$$

$$u = u_n^{\complement V} + (u - u_n^{\complement V}) \leq v + w.$$

Since $u = w$ in $U \setminus V$, it follows that $u \leq v + w$ in all of U, that is, $f \leq w$, and hence

$$R_f \leq w \qquad (\text{in } U).$$

In particular, we obtain in V

$$w = R_f \leq (R_f)^{\complement V} + u - u_n^{\complement V},$$

$$R_f + u_n^{\complement V} \leq (R_f)^{\complement V} + u,$$

and consequently, letting $n \to \infty$,

$$R_f + u^{\complement V} \leq (R_f)^{\complement V} + u \qquad (\text{in } V). \tag{32}$$

Consider the function $g \geq 0$ on U defined by

$$g(x) = \begin{cases} u(x) - R_f(x) & \text{when} \quad R_f(x) < +\infty, \\ +\infty & \text{when} \quad R_f(x) = +\infty. \end{cases}$$

Clearly, g is finely continuous in the finely open set $[R_f < +\infty]$, hence finely u.s.c in all of U. Moreover,

$$u = R_f + g \qquad \text{everywhere in } U.$$

Hence $u^{\complement V} = (R_f)^{\complement V} + g^{\complement V}$. It now follows from (32) that g is nearly finely hyperharmonic in U (Def. 11.1). With V as above, we have, in fact,

$$g^{\complement V} = u^{\complement V} - (R_f)^{\complement V} \leq u - R_f = g$$

at any point of U at which $g < +\infty$ (and hence also $(R_f)^{\complement V} \leq R_f < +\infty$).

By the fine continuity of g in $[R_f < +\infty]$ we obtain $\hat{g} = g$ in this finely open set. It follows that $u = R_f + \hat{g}$ in $[R_f < +\infty]$, and trivially in $[R_f = +\infty]$, hence in all of U. This shows that indeed $R_f \preccurlyeq u$. ▯

11.15. In view of the above lemma it has now been verified altogether that the cone $\mathcal{W}$ of all finely hyperharmonic functions on the given finely open set U satisfies the axioms set down in [22 bis, Chap. 4] when U is given the fine topology. We proceed to list a number of consequences of this circumstance, cf. [22 bis].

a) $\mathcal{W}$ is an upper complete lattice also with respect to the <u>specific</u> order. Hence $\mathcal{W}$ is conditionally complete in this order, and $\mathcal{W}^+ = \mathcal{U}(U)$ is specifically complete.

b) For any family of functions $u_i \in \mathcal{U}(U)$ the specific supremum $\curlyvee u_i$ and the pointwise supremum $\vee u_i$ (relative to $\mathcal{U}(U)$) are related as follows:

$$\curlyvee u_i \succcurlyeq \vee u_i$$

with equality if (u_i) is <u>specifically upper directed</u>, in which case we thus have

$$\curlyvee u_i = \vee u_i = \sup u_i .$$

c) For any family of functions $u_i \in \mathcal{U}(U)$ we have $\curlywedge u_i \preccurlyeq \wedge u_i$ (this is easy). If (u_i) is specifically lower directed, then

$$\curlywedge u_i = \wedge u_i \ \left(= \widehat{\inf u_i}\right);$$

and moreover, for any $x \in U$ such that $u_i(x) < +\infty$ for some i,

$$(\curlywedge u_i)(x) = (\wedge u_i)(x) = \inf u_i(x) .$$

d) Let $u \in \mathcal{U}(U)$, and let $A \subset U$. If $u < +\infty$ in A, then

$$R_u^A = \inf \{ R_u^V \mid V \text{ finely open, } A \subset V \subset [u < +\infty] \}.$$

e) For any $u, v \in \mathcal{U}(U)$ and any set $A \subset U$,

$$R_{u+v}^A = R_u^A + R_v^A \; ; \qquad \hat{R}_{u+v}^A = \hat{R}_u^A + \hat{R}_v^A .$$

f) For any $u \in \mathcal{U}(U)$ and any sets $A, B \subset U$,

$$R_u^{A \cup B} + R_u^{A \cap B} \leq R_u^A + R_u^B ,$$

$$\hat{R}_u^{A \cup B} + \hat{R}_u^{A \cap B} \leq \hat{R}_u^A + \hat{R}_u^B ,$$

$$\hat{R}_u^{A \cup B} \preccurlyeq \hat{R}_u^A + \hat{R}_u^B .$$

11.16. On the cone $\mathcal{S}(U)$ of all finely superharmonic functions ≥ 0 (Def. 10.4) in the given finely open set U we have the induced specific order $\preccurlyeq$ defined by

$$s_1 \preccurlyeq s_2 \iff \exists\, s \in \mathcal{S}(U) : \; s_1 + s = s_2 .$$

In this situation, s is uniquely determined by s_1 and s_2. (In fact, $s = s_2 - s_1$ holds pointwise q.e., viz. everywhere in $[s_1 < +\infty]$, whence the uniqueness of s by fine continuity.)

It is therefore clear that $\mathcal{S}(U)$, endowed with the specific order, is a pre vector lattice in the sense of [22 bis, §8.1], in other words that $\mathcal{S}(U)$ can be imbedded as the positive cone in a vector lattice (Riesz space). Moreover, this vector lattice is conditionally complete because the lattice $\mathcal{S}(U)$ is lower complete. In fact, $\mathcal{S}(U)$ is a face (= hereditary convex subcone) of the complete lattice cone $\mathcal{U}(U)$ with the specific order, cf. a) of §11.15.

Consequently, the usual rules concerning the specific order for superharmonic functions ≥ 0 (see §1.4 - §1.6) carry over to the present, more general case in view of [22 bis, Prop. 8.1.1].

For these rules it was assumed, in §1.4 - §1.6, that certain of the hyperharmonic functions ≥ 0 in question were superharmonic. In the present frame of a strong harmonic space satisfying axiom (D) such limitations are easily shown to be superfluous because a hyperharmonic function in a connected open set is either identically $+\infty$ or else superharmonic. Invoking Theorem 12.9 below, together with the local connectivity of the fine topology, it is easily shown, in exactly the same way, that the corresponding rules concerning the specific order on the cone $\mathcal{S}(U)$ actually hold for the larger cone $\mathcal{U}(U)$ of all finely hyperharmonic functions ≥ 0 on the given finely open set U. In particular, $\mathcal{U}(U)$ has the Riesz decomposition property also with respect to the specific order.

The fine potentials (Def. 10.5) relative to U form a band in $\mathcal{S}(U)$ with the specific order in view of §10.6. The orthogonal band consists of all finely harmonic functions ≥ 0 (Theorem 10.7).

The stable fine potentials (Def. 10.9) relative to U likewise form a band in $\mathcal{S}(U)$ in view of Cor. 10.9. Its orthogonal band within $\mathcal{S}(U)$ consists of all functions in $\mathcal{S}(U)$ which are finely harmonic quasi everywhere (Theorem 10.10).

We shall denote by $\mathcal{P}^s(U)$ the band of all stable fine potentials relative to U. It is itself a lower complete pre vector lattice. (Recall that, according to Theorem 10.12, $\mathcal{P}^s(\Omega) = \mathcal{P}^s$, the band of semibounded, ordinary potentials on Ω within $\mathcal{S}(\Omega) = \mathcal{S}$, the usual superharmonic functions ≥ 0 on Ω. In particular, we have thus determined the band orthogonal to $\mathcal{P}^s$ within $\mathcal{S}$ with the usual specific order, viz. all $s \in \mathcal{S}$ with s finely harmonic q.e.)

11.17. For any numerical function p on the given finely open set U we denote by $S_q(p)$ the quasi harmonic support of p relative to U, that is, the complement within U of the largest finely open subset of U in which p is finely harmonic q.e. (cf. Def. 8.8 for $U = \Omega$). Clearly $S_q(U)$ is finely closed relatively to U, and in fact a base relative to U in the sense that

$$U \cap b(S_q(p)) = S_q(p) .$$

Recall that the face in $\mathcal{S}(U)$ formed by all finite valued, fine potentials relative to U is contained in $\mathcal{P}^s(U)$, and that $S_q(p)$ equals the fine harmonic support $S_f(p)$ relative to U for any such finite fine potential on account of Cor. 9.15.

Theorem. For any finely open set U the quasi harmonic support S_q (relative to U) is an abstract carrier when considered as a mapping $p \longmapsto S_q(p)$ of $\mathcal{P}^s(U)$ into the class of all relatively finely closed subsets of U.

Proof. The concept of an abstract carrier referred to here is that of Constantinescu and Cornea [22 bis, §8.1].

Since the fine topology on U is regular, the assertion of the theorem therefore amounts to the following 3 properties of S_q (relative to U), applied to stable fine potentials $p \in \mathcal{P}^s(U)$:

a) $p = 0 \quad \Longleftrightarrow \quad S_q(p) = \emptyset$.

b) $p_1 \preccurlyeq p_2 \quad \Longrightarrow \quad S_q(p_1) \subset S_q(p_2)$.

c) For any $p \in \mathcal{P}^s(U)$ and any two sets $A_1, A_2 \subset U$ which cover U and are finely closed relatively to U, there exist $p_1, p_2 \in \mathcal{P}^s(U)$ such that $p = p_1 + p_2$ and $S_q(p_i) \subset A_i$ ($i = 1, 2$).

Ad a). $S_q(p) = \emptyset$ means that p is finely harmonic q.e. in U, or equivalently, by Lemma 10.9, that $p = 0$.

Ad b). When $p_1 \preccurlyeq p_2$, p_1 is obviously finely harmonic in any finely open set in which p_2 is finely harmonic, thus in particular in $U \setminus S_q(p_2)$ off some polar set.

Ad c). Using the last result in f) of §11.15, we obtain

$$p = \hat{R}_p^U = \hat{R}_p^{A_1 \cup A_2} \preccurlyeq \hat{R}_p^{A_1} + \hat{R}_p^{A_2}.$$

Here $\hat{R}_p^{A_i} \leqslant p_i$, and so $\hat{R}_p^{A_i} \in \mathcal{P}^s(U)$ ($i = 1, 2$). By the Riesz decomposition property of $\mathcal{P}^s(U)$ (or of $\mathcal{S}(U)$) in the specific order, there exist $p_1, p_2 \in \mathcal{P}^s(U)$ such that $p = p_1 + p_2$ and that $p_i \preccurlyeq \hat{R}_p^{A_i}$ ($i = 1, 2$). Invoking b) above, together with

Cor. 11.13, we conclude that, for $i = 1, 2$,

$$S_q(p_i) \subset S_q(\hat{R}_p^{A_i}) \subset A_i . \quad \Box$$

11.18. Theorem. The band $\mathcal{P}^{\wedge}(U)$ of all stable, fine potentials relative to U is generated by the finite valued, fine potentials relative to U.

Proof. Since every finite, fine potential is stable (Cor. 10.9), it remains to prove that every stable, fine potential p is the specific supremum of its finite valued specific minorants. (Equivalently, by the quasi Lindelöf principle, p is representable as the sum of a sequence of finite, fine potentials.)

Choose an ordinary locally bounded potential $r > 0$ on Ω, and consider the relatively finely closed subsets of U:

$$A_n := [p \leqslant nr] \; ; \qquad B_n := [p \geqslant nr].$$

Since $A_n \cup B_n = U$, it follows from the preceding theorem that there exist fine potentials p_n and q_n relative to U such that $p = p_n + q_n$ and

$$S_q(p_n) \subset A_n \; ; \qquad S_q(q_n) \subset B_n .$$

According to the relative fine domination principle (Cor. 10.11), $p_n \leqslant nr < +\infty$ in all of U because nr is finely hyperharmonic in U. Now write

$$q := \bigwedge_n q_n .$$

Since q_n is finely harmonic q.e. in $U \setminus B_n$, the same applies to q, and so q is finely harmonic q.e. in U because $\bigcap B_n = [p = +\infty]$ is polar. Invoking Lemma 10.9, we conclude that $q = 0$, $p = \bigvee p_n$. $\Box$

11.19. Let $\mathcal{C}^+(U)$ denote the lattice cone of all bounded, finely continuous functions ≥ 0 on the given finely open set U.

On account of Theorem 11.17 we obtain by application of [22 bis, §8.1] a unique mapping

$$(f, p) \mapsto f \cdot p$$

of $\mathcal{C}^+(U) \times \mathcal{P}^s(U)$ into $\mathcal{P}^s(U)$ which is affine in the first variable f, and such that

$$1 \cdot p = p \; ; \qquad S_q(f \cdot p) \subset [f \neq 0]^\sim$$

for all $f \in \mathcal{C}^+(U)$, $p \in \mathcal{P}^s(U)$. (Note that $[f \neq 0]^\sim$ is the fine support of f.)

This mapping is called the specific multiplication of functions $f \in \mathcal{C}^+(U)$ by stable, fine potentials $p \in \mathcal{P}^s(U)$. It is affine also in the second variable p, hence increasing in each variable (using the specific order on $\mathcal{P}^s(U)$). The specific multiplication possesses the following further properties, in which sup and inf refer to the complete lattice

$$\mathcal{F}^+(U) := [0, +\infty]^U,$$

and $\curlyvee$, $\curlywedge$ to the complete lattice $\mathcal{U}(U)$ of all finely hyperharmonic functions ≥ 0 (with the specific order):

a) $\curlyvee_i (f_i \cdot p) = (\sup_i f_i) \cdot p \quad ; \quad \curlywedge_i (f_i \cdot p) = (\inf_i f_i) \cdot p$,

for any $p \in \mathcal{P}^s(U)$ and any family (f_i) on $\mathcal{C}^+(U)$ such that $\sup_i f_i \in \mathcal{C}^+(U)$, resp. $\inf_i f_i \in \mathcal{C}^+(U)$.

b) $\bigvee_i (f\cdot p_i) = f\cdot \bigvee_i p_i$; $\bigwedge_i (f\cdot p_i) = f\cdot \bigwedge_i p_i$,
for any $f \in \mathscr{C}^+(U)$ and any family (p_i) on $\mathscr{P}^\Delta(U)$ such that (in the former case) $\bigvee p_i \in \mathscr{P}^\Delta(U)$.

c) $(f_1 f_2)\cdot p = f_1\cdot(f_2\cdot p)$
for any $f_1, f_2 \in \mathscr{C}^+(U)$, $p \in \mathscr{P}^\Delta(U)$.

d) $S_q(f\cdot p) \subset [f \neq 0]^\sim \cap S_q(p)$
for any $f \in \mathscr{C}^+(U)$, $p \in \mathscr{P}^s(U)$.

Although the fine topology in U is not locally compact, there is no difficulty in extending the affine mapping $f \mapsto f\cdot p$ for fixed $p \in \mathscr{P}^\Delta(U)$ so as to be defined for more general functions $f \geq 0$ on U in a suitable way, using well-known ideas from topological integration theory, and profiting from the complete regularity of the fine topology in U .

11.20. First we extend the mapping from $\mathscr{C}^+(U)$ to the lattice cones

$$\mathcal{G}(U) = \{ f \in \mathcal{F}^+(U) \mid f \text{ is finely l.s.c.} \} ,$$

$$\mathcal{H}(U) = \{ f \in \mathcal{F}^+(U) \mid f \text{ is finely u.s.c. and bounded} \} ,$$

by the following definitions with $f \in \mathcal{G}(U)$ and $f \in \mathcal{H}(U)$, respectively:

$$f\cdot p = \bigvee \{ \varphi\cdot p \mid \varphi \in \mathscr{C}^+(U), \ \varphi \leq f \},$$

$$f\cdot p = \bigwedge \{ \varphi\cdot p \mid \varphi \in \mathscr{C}^+(U), \ \varphi \geq f \}.$$

Clearly $f\cdot p \in \mathscr{P}^\Delta(U)$ if $f \in \mathcal{H}(U)$. For $f \in \mathcal{G}(U)$ we have $f\cdot p \in \mathcal{U}(U)$, and so $f\cdot p \in \mathscr{P}^\Delta(U)$ if and only if $f\cdot p < +\infty$ quasi everywhere.

The extension to $\mathcal{H}(U)$ is affine and increasing in each variable and satisfies c) and d) together with the latter part of a) and b) (with $\mathcal{H}(U)$ in place of $\mathcal{C}^+(U)$ throughout).

We shall limit ourselves to giving the proof of d) in the case $f \in \mathcal{H}(U)$. Consider any function $\psi \in \mathcal{C}^+(U)$ with $0 \leq \psi \leq 1$ and such that $\psi = 1$ on the fine support $[f \neq 0]^\sim$. For any φ as in the above definition of $f \cdot p$ for $f \in \mathcal{H}(U)$, $\varphi\psi$ has the same properties as φ, and since $\varphi\psi \leq \varphi$, we get

$$f \cdot p = \bigwedge \{ (\varphi\psi) \cdot p \mid \varphi \in \mathcal{C}^+(U), \ \varphi \geq f \},$$

$$S_q(f \cdot p) \subset S_q(\varphi\psi \cdot p) \subset [\psi \neq 0]^\sim \cap S_q(p).$$

From this follows d) because the finely closed set $[f \neq 0]^\sim$, by complete regularity, is the intersection of the sets $[\psi \neq 0]^\sim$ as ψ ranges over the functions specified above.

The extension to $\mathcal{G}(U)$ is likewise affine and increasing in either variable. It satisfies the former part of a) and b) (with $\mathcal{G}(U)$ in place of $\mathcal{C}^+(U)$). Moreover, c) and d) carry over with the understanding that $f_2 \cdot p$ and $f \cdot p$, respectively, is supposed to be of class $\mathcal{P}^s(U)$, that is, to be finite q.e.

Finally, it is easily verified that

$$g \cdot p \geq h \cdot p$$

for any $g \in \mathcal{G}(U)$, $h \in \mathcal{H}(U)$ with $g \geq h$.

11.21. We are now prepared to complete the extension procedure. For any numerical function $f \geq 0$ on U define "upper and lower integrals" as follows:

$$(f\cdot p)^* = \curlywedge\{g\cdot p \mid g\in\mathcal{G}(U),\ g\geq f\},$$

$$(f\cdot p)^* = \curlyvee\{h\cdot p \mid h\in\mathcal{H}(U),\ h\leq f\}.$$

Like in §11.20, $\curlywedge$ and $\curlyvee$ refer to the specific order on $\mathcal{U}(U)$, but we might equally well write $\bigwedge$ and $\bigvee$, respectively, thus referring to the pointwise order relative to $\mathcal{U}(U)$, because $g\cdot p$ and $h\cdot p$ range over specifically lower, resp. specifically upper, directed subsets of $\mathcal{U}(U)$, cf. §11.15.

The mapping $(f,p)\longmapsto(f\cdot p)^*$ (resp. $(f\cdot p)_*$) of $\mathcal{F}^+(U)\times\mathcal{P}^{\Delta}(U)$ into $\mathcal{U}(U)$ is increasing and positive homogeneous in each variable, and additive in the latter variable p . It is furthermore countably subadditive (resp. countably superadditive) in the former variable f . Also the former (resp. latter) part of a), §11.19, carries over for countable families (f_ι) on $\mathcal{F}^+(U)$ in the case of the "upper" (resp. "lower") "integral". Moreover, d) of §11.19 carries over as follows

$$S_q((f\cdot p)^*) \subset b([f\neq 0])\cap S_q(p) \quad \text{if} \quad (f\cdot p)^* < +\infty \text{ q.e.,}$$

$$S_q((f\cdot p)_*) \subset b([f\neq 0])\cap S_q(p) \quad \text{if} \quad (f\cdot p)_* < +\infty \text{ q.e.}$$

(Cf. the corresponding proof in §11.20 as to the non trivial former case.) – We have here replaced the fine support $[f\neq 0]^\sim$ by its base, which is permitted since, for any function u on U, the quasi support $S_q(u)$ is a base relative to U .

Note that, in particular, $(f\cdot p)^* = 0$ for any function $f\in\mathcal{F}^+(U)$ such that $f = 0$ q.e. In view of the subadditivity in the former variable, this implies that, for $f_1, f_2 \in \mathcal{F}^+(U)$,

$$f_1 = f_2 \quad \text{q.e.} \Longrightarrow S_g((f_1\cdot p)^*) = S_g((f_2\cdot p)^*) . \quad (33)$$

We omit the proofs of the above assertions concerning the algebraic and order properties of the extensions $(f\cdot p)^*$ and $(f\cdot p)_*$ since they run along well-known lines from the theory of the Daniell integral. Again, b) and c) of §11.15 allow us to pass instead to pointwise statements in the case of directed families.

It follows from the last result of §11.20 that

$$(f\cdot p)_* \leq (f\cdot p)^*$$

for every $f \in \mathcal{F}^+(U)$. The sign of equality occurs, e.g., for f of class $\mathcal{G}(U)$ or $\mathcal{H}(U)$. Whenever the equality sign prevails we may therefore consistently denote the common value by $f\cdot p$ $(\in \mathcal{U}(U))$.

For any two functions $f_1 , f_2 \in \mathcal{F}^+(U)$ we have

$$((f_1 + f_2)\cdot p)_* \leq (f_1\cdot p)_* + (f_2\cdot p)^* \leq ((f_1+f_2)\cdot p)^*.$$

As to the former inequality, let $h \in \mathcal{H}(U)$, $h \leq f_1 + f_2$, and let $g \in \mathcal{G}(U)$, $g \geq f_2$. Then $h \leq f_1 + g$, hence $(h-g)^+ \in \mathcal{H}(U)$, $(h-g)^+ \leq f_1$, and so

$$h\cdot p \leq (h-g)^+\cdot p + g\cdot p \leq (f_1\cdot p)_* + g\cdot p$$

by the subadditivity of the "upper integral" together with the fact that h , $(h-g)^+$, and g all belong to $\mathcal{H}(U)$ or $\mathcal{G}(U)$. - The latter inequality above is obtained similarly, invoking again the last result of §11.20.

<u>Definition</u>. Let $p \in \mathcal{P}^{\wedge}(U)$ be given. A function $f \in \mathcal{F}^+(U)$ is called p-<u>integrable</u>, and we write $f \in \mathcal{L}(p)$, if

$$(f \cdot p)_* = (f \cdot p)^* \quad \overset{q.e.}{\lesssim} + \infty \quad .$$

A function $f \in \mathcal{F}^+(U)$ is called p-<u>measurable</u>, and we write $f \in \mathcal{M}(p)$, if $\inf(f, \lambda)$ is p-integrable for every finite constant λ .

Clearly, $\mathcal{H}(U) \subset \mathcal{L}(p)$, $\mathcal{G}(U) \subset \mathcal{M}(p)$. In particular, every bounded p-measurable function is p-integrable. It is easily shown that

$$(f \cdot p)_* = (f \cdot p)^* \quad \text{for every } f \in \mathcal{M}(p) ,$$

and so we may write simply $f \cdot p$ for such f . (Consider, in fact, the increasing sequence of functions $\inf(f, n) \in \mathcal{L}(p)$.)

<u>Theorem</u>. $\mathcal{M}(p)$ is a convex subcone of $\mathcal{F}^+(U)$, stable under countable supremum or infimum (relative to $\mathcal{F}^+(U)$). $\mathcal{L}(p)$ is a face in $\mathcal{M}(p)$, hence stable under countable infimum and finite supremum. For any f_1 , $f_2 \in \mathcal{M}(p)$ (resp. $\mathcal{L}(p)$) we have $(f_1 - f_2)^+ \in \mathcal{M}(p)$ (resp. $\mathcal{L}(p)$).

The mapping $f \longmapsto f \cdot p$ of $\mathcal{M}(p)$ into $\mathcal{U}(U)$ is affine, hence increasing, and has the following further properties involving countable families of functions $f_n \in \mathcal{M}(p)$:

$$(\sum_n f_n) \cdot p = \sum_n (f_n \cdot p) ,$$

$$\bigvee_n (f_n \cdot p) = (\sup_n f_n) \cdot p ,$$

$$\bigwedge_n (f_n \cdot p) = (\inf_n f_n) \cdot p \quad \text{provided that } f_n \in \mathcal{L}(p).$$

Finally

$$S_q(f \cdot p) \subset b([f \neq 0]) \cap S_q(p) \quad \text{for any } f \in \mathcal{L}(p).$$

Proof. All these assertions are immediate consequences of the properties of $(f\cdot p)^*$ and $(f\cdot p)_*$ listed above. ▯

In view of b) and c) of §11.15 we furthermore obtain the following pointwise properties:

$$\sup_n (f_n \cdot p) = (\sup_n f_n)\cdot p$$

for any increasing sequence (f_n) on $\mathcal{M}(p)$; and

$$\inf_n [(f_n \cdot p)(x)] = [(\inf_n f_n)\cdot p](x)$$

for any decreasing sequence (f_n) on $\mathcal{L}(p)$ and any point $x \in U$ such that $(f_n\cdot p)(x) < +\infty$.

For any set $A \subset U$ which is p-measurable in the sense that $1_A \in \mathcal{M}(p)$ (hence $1_A \in \mathcal{L}(p)$), the finely stable potential

$$p_A := 1_A \cdot p \in \mathcal{P}^s(U)$$

is called the specific restriction of p to A . Note that p_A does not change if A is modified by a polar set. This appears from the general observation (33), or directly from the relation

$$S_q(p_A) \subset b(A) \cap S_q(p)$$

according to which every polar set is "p-negligible".

11.22. Let $\mathcal{E}(U)$ denote the smallest subclass of $\mathcal{F}^+(U)$ which is stable under countable supremum and infimum, and which contains $\mathcal{C}^+(U)$ (the bounded, finely continuous functions ≥ 0 on U). Then

$$\mathcal{E}(U) \subset \mathcal{M}(p) \qquad \text{for every } p \in \mathcal{P}^s(U) .$$

For any given $p \in \mathcal{P}^s(U)$ the mapping $f \longmapsto f\cdot p$ of $\mathcal{E}(U)$ into $\mathcal{U}(U)$ is uniquely determined as an affine mapping $\varphi : \mathcal{E}(U) \longrightarrow \mathcal{U}(U)$ with the properties $\varphi(1) = p$, $S_q(\varphi(f)) \subset [f \neq 0]^\sim$ for all $f \in \mathcal{E}(U)$, and

$$\varphi(\sup_n f_n) = \bigvee_n \varphi(f_n)$$

for every (say increasing) sequence of functions $f_n \in \mathcal{E}(U)$. (This follows easily from the uniqueness of the original mapping $f \longmapsto f\cdot p$ of $\mathcal{C}^+(U)$ into $\mathcal{P}^s(U)$.)

Using this uniqueness result, one may establish the following rules concerning $f\cdot p$ as a function of $p \in \mathcal{P}^s(U)$ for fixed $f \in \mathcal{E}(U)$. Let (p_n) denote a countable family of stable fine potentials relative to U. Then:

$$f\cdot\sum_n p_n = \sum_n (f\cdot p_n) \qquad \text{if } \sum_n p_n \in \mathcal{P}^s(U)\,,$$

$$f\cdot\bigvee_n p_n = \bigvee_n (f\cdot p_n) \qquad \text{if } \bigvee_n p_n \in \mathcal{P}^s(U)\,,$$

$$f\cdot\bigwedge_n p_n = \bigwedge_n (f\cdot p_n) \qquad \text{if each } (f\cdot p_n) \in \mathcal{P}^s(U)\,.$$

(The second property is obtained by defining $\varphi(f) = \bigvee (f\cdot p_n)$ and writing $p = \bigvee p_n$. The remaining two properties are simple consequences.)

The last two properties above in turn lead to the following properties for the upper, resp. lower "integral":

$$(f\cdot\bigvee_n p_n)^* = \bigvee_n (f\cdot p_n)^* \qquad \text{if } \bigvee_n p_n \in \mathcal{P}^s(U)\,,$$

$$(f\cdot\bigwedge_n p_n)_* = \bigwedge_n (f\cdot p_n)_* \qquad \text{if each } (f\cdot p_n)_* \in \mathcal{P}^s(U)\,.$$

The proof of these latter results is straightforward in the case of finite families (p_n), and the same proof works for countable families in view of (33), p. 141, because $\inf_n g_n$, resp. $\sup_n h_n$, differs only in a polar set from a function of class $\mathcal{E}(U)$ whenever (g_n) , resp. (h_n) , is a sequence of functions of class $\mathcal{G}(U)$, resp. $\mathcal{H}(U)$. This follows from the quasi Lindelöf principle.

CHAPTER IV

Applications

12. Properties involving fine connectivity and balayage of measures

12.1. We have established above (as corollaries to Theorems 9.11 and 9.8) that our strong harmonic space Ω with axiom (D) is finely locally connected, and that every subdomain of Ω in the initial topology is likewise a fine domain (= finely connected finely open set).

As a consequence of the local connectivity of the fine topology, the fine components of a finely open set are finely open, and their number is hence at most countable in view of the quasi Lindelöf principle (any finely open polar set being void).

12.2. Theorem. Let U denote a finely open set and e a polar set. Then $U \setminus e$ is finely connected if and only if U is finely connected.

Proof. If $U \setminus e$ is finely connected then so is U since $U \setminus e$ is finely dense in U. Conversely, suppose that U is finely connected, that is, a fine domain. Choose $p \in \mathcal{P}^a$ so that $p > 0$, and note that p is finely hyperharmonic in every finely open set by Theorem 8.7 and the sheaf property (§8.6). Let V_0 and V_1 denote two finely open sets such that $V_0 \cup V_1 = U \setminus e$ and $V_0 \cap V_1 = \emptyset$. The function u defined in $U \setminus e$ by $u = ip$ in V_i $(i = 0, 1)$ is finely hyperharmonic and $\geqslant 0$, hence admits a finely hyperharmonic extension to U according to Theorem 9.14. This extension u is finely continuous by Theorem 9.10, and so is therefore u/p. In $U \setminus e$, and hence in all of U, u/p

takes the values 0 and 1 only. Since U is finely connected, we conclude that either $u \equiv 0$ or $u \equiv p$, whence either V_1 or V_0 is void. ▯

This theorem and some of its consequences below were obtained in [27] in the case of a Green space.

12.3. Lemma. For any set $A \subset \Omega$ we have

$$i(\partial_f A) = i(A) \cup i(\complement A),$$

$$b(\partial_f A) = b(A) \cap b(\complement A),$$

and hence $\mu^A = \mu^{\partial_f A}$ for every admissible measure μ carried by $b(\complement A)$.

Proof. It was noted in §3.7 that $i(A) \subset i(\partial_f A)$. Hence also $i(\complement A) \subset i(\partial_f A)$. Conversely, let $x \in i(\partial_f A)$, that is, x is a polar and finely isolated point of $\partial_f A$. By the local connectivity of the fine topology, there exists a fine domain W such that $W \cap \partial_f A = \{x\}$, and hence $(W \setminus \{x\}) \cap \partial_f A = \emptyset$. Since x is polar, $W \setminus \{x\}$ is finely connected by the preceding theorem. Supposing, e.g., that $x \in A$, we conclude that $W \setminus \{x\} \subset \complement A$ (hence $x \in i(A)$), the alternative $W \setminus \{x\} \subset A$ being impossible since it would imply that $W \subset A$, in contradiction with $x \in \partial_f A$.

Next we obtain

$$\begin{aligned} b(\partial_f A) &= \partial_f A \setminus i(\partial_f A) \\ &= \tilde{A} \cap (\complement A)^{\sim} \cap \complement\, i(A) \cap \complement\, i(\complement A) \\ &= b(A) \cap b(\complement A). \end{aligned}$$

For any μ as stated, μ^A is carried by $b(A) \cap b(\complement A) = b(\partial_f A)$ (§4.8), whence $\mu^{\partial_f A} = \mu^A$ by [27, Lemme 3.1]. ▯

12.4. Lemma. For any finely closed set A we have

$$i(\partial_f A) = \partial_f\, i(A) = i(A) \quad (\subset \partial_f A),$$

$$b(\partial_f A) = \partial_f\, b(A) = b(A) \cap \partial_f A .$$

In particular, A (finely closed) is a base if and only if $\partial_f A$ is a base.

Proof. The partition $A = b(A) \cup i(A)$ of the finely closed set A into disjoint, finely closed sets $b(A)$ and $i(A)$ (§3.6) yields

$$\partial_f A = \partial_f\, b(A) \cup \partial_f\, i(A) = \partial_f\, b(A) \cup i(A)$$

because $i(A)$ is polar (cf. §4.2 and §4.3). Taking bases, we get from the preceding lemma applied to $b(A)$ in place of A, and noting that $\complement b(A)$ is finely open (see at this point §3.7),

$$b(\partial_f A) = b(\partial_f\, b(A)) = b(b(A)) \cap b(\complement b(A)) = \partial_f\, b(A).$$

Clearly, $b(\partial_f A) \subset b(A) \cap \partial_f A$ since $\partial_f A$ is finely closed and contained in A . The opposite inclusion follows again from Lemma 12.3 because $\complement A$ is finely open, and so, by §3.7,

$$\partial_f A = \partial_f(\complement A) \subset (\complement A)^{\sim} = b(\complement A).$$

Finally, $i(\partial_f A) = i(A)$ follows from Lemma 12.3, now applied to $\complement A$, noting that $i(\complement A) = \emptyset$ since $\complement A$ is finely open. ▯

12.5. Applying Lemma 12.4 to the complement of a finely open set U , we propose the following definition.

<u>Definition</u>. Let U denote a finely open set. The points of

$$b(\partial_f U) = b(\complement U) \cap \partial_f U$$

are called the <u>regular</u> points of $\partial_f U$, or the regular (fine boundary) points for U . The remaining points of $\partial_f U$, forming the polar set

$$i(\partial_f U) = i(\complement U)$$

are called the <u>irregular</u> points of $\partial_f U$, or the irregular (fine boundary) points for U .

Thus the irregular points are precisely those points of $\complement U$ at which $\complement U$ is thin. In particular, this notion of an irregular point coincides with the usual concept of an irregular point for U in the special case where U is open in the initial topology.

A finely open set U without irregular points is the same as a <u>regular</u> finely open set in the sense of §4.3, that is, the complement of a base, or equivalently, by Lemma 12.4, a finely open set whose fine boundary is a base.

For any fine component W of a finely open set U , a fine boundary point $x \in \partial_f W$ is regular for W if and only if x is regular for U (because $\complement W \supset \complement U$, whereas $\partial_f W \subset \partial_f U$).

In particular, any fine component of a regular, finely open set is regular. (The converse is false.)

In view of §4.3 it follows that the <u>regular fine domains</u> form a basis for the fine topology on Ω .

For any finely open set U the set

$$V := \complement b(\complement U) = U \cup i(\complement U)$$

obtained by adding to U the polar set $i(\complement U)$ of all irregular points for U, is a regular finely open set, and of course the smallest such set containing U. Clearly,

$$\partial_f U = \partial_f V \cup i(\complement U); \quad \partial_f V = b(\partial_f U); \quad \tilde{V} = \tilde{U}.$$

According to Theorem 12.2, V is finely connected if and only if U is finely connected.

Note that $V = U \cup i(\complement U)$ is not in general open in the initial topology when U is open. This fact contributes to make the initial topology less adapted than the fine topology to the study of the Dirichlet problem in the irregular case (see also §14).

12.6. Theorem. Let $u \geq 0$ be finely hyperharmonic in a fine domain U. Then either $u > 0$ or $u \equiv 0$.

Proof. The function

$$\sup_n nu = +\infty \cdot 1_E \ , \quad \text{where} \quad E := \{x \in U \mid u(x) > 0\}$$

is finely hyperharmonic by Cor. 9.5, and hence finely continuous in U by Theorem 9.10. Therefore E is finely open and finely closed relatively to U, and consequently either $E = U$ or $E = \emptyset$. ▮

Corollary. For any regular fine domain U the "harmonic measures" $\varepsilon_x^{\complement U}$, $x \in U$, have all the same null sets.

In fact, for any set $E \subset \Omega$, the function u defined by

$$u(x) = (\varepsilon_x^{\complement U})^*(E) = \int^* 1_E \, d\varepsilon_x^{\complement U}$$

is finely hyperharmonic in U according to Theorem 9.13, whence the result by the above theorem.

<u>Remark</u>. Unlike the situation in the case of a usual regular domain, the <u>integrability</u> of a function f (say ≥ 0) with respect to $\varepsilon_x^{\complement U}$ may depend on $x \in U$ (when U is a regular fine domain) because the finely hyperharmonic function u defined by $u(x) = \int^* f \, d\varepsilon_x^{\complement U}$ may take the value $+\infty$ for all x in some set $E \subset U$ without being identically infinite in U. (Such a set E is, however, necessarily <u>polar</u> according to Theorem 12.9 below.) See Ex. 8.10 (taking $f = G\varepsilon_x$, $E = \{x\}$).

12.7. <u>Theorem</u>. Let μ be an admissible measure on Ω, and U a regular finely open set (the complement of a base B). Let U^μ denote the union of those fine components V of U for which $\mu(V) > 0$. Then $\mu(U \setminus U^\mu) = 0$, and $\partial_f U^\mu \subset \partial_f U$. Moreover,

a) $\complement U^\mu$ (also written B_μ) is the largest among all bases E such that $\mu^E = \mu^B$.

b) If U carries μ then the fine support of μ^B is

$$\operatorname{supp}_f \mu^B = \partial_f U^\mu = \partial_f B_\mu .$$

<u>Proof</u>. Clearly, $U \setminus U^\mu$ is finely open, and μ-negligible since U has only countably many fine components. Also $\partial_f U^\mu \subset \partial_f U$.

Ad a). It was proved in [27, Théorème 3 a)] that there exists a largest base B_μ among the bases $E \supset B$ with $\mu^E = \mu^B$.

For <u>any</u> base E with $\mu^E = \mu^B$ we obtain $E \subset B_\mu$ as follows: According to Constantinescu [21, Theorem 3.1], $\mu^{E\cup B} \leqslant \sup(\mu^E, \mu^B) = \mu^B$, and so

$$\int \hat{R}_u^{E\cup B} d\mu \leqslant \int \hat{R}_u^B d\mu$$

for every $u \in \mathcal{U}$. The converse inequality being trivial, we conclude that $\mu^{E\cup B} = \mu^B$, and hence $E \subset E\cup B \subset B_\mu$ since $E\cup B \supset B$.

The inclusion $\complement B_\mu \subset U^\mu$ was obtained in [27, Théorème 3 b)] according to which every fine component of $\complement B_\mu$ is non negligible with respect to μ, and hence extends to a non negligible fine component V of $U = \complement B$ ($\supset \complement B_\mu$), that is, $V \subset U^\mu$.

To prove that actually $\complement B_\mu = U^\mu$, choose a <u>strict</u> <u>potential</u> $p \in \mathcal{P}^c$ in the sense of Constantinescu with $\int p\,d\mu < +\infty$ (see §3.4 above). From $\mu^{B_\mu} = \mu^B$ follows

$$\int p^{B_\mu} d\mu = \int p\,d\mu^{B_\mu} = \int p\,d\mu^B = \int p^B d\mu$$

that is, $\int u\,d\mu = 0$, when u is defined by

$$u = p^{B_\mu} - p^B = \hat{R}_p^{B_\mu} - \hat{R}_p^B .$$

Note that $u \geqslant 0$ in Ω since $B_\mu \supset B$. Moreover, u is finely hyperharmonic in $\complement B$ because p^B is finely harmonic there according to Lemma 9.3.

Now consider any fine component V of U^μ, that is, any fine component V of U such that $\mu(V) > 0$. By the local connectivity of the fine topology, V is a fine domain, and hence it follows from the preceding theorem that $u \equiv 0$ in V, the alternative $u > 0$ in V being ruled out since $\mu(V) > 0$ and $\int u\,d\mu = 0$. Consequently, $u \equiv 0$ in U^μ, that is

$$p^{B_\mu} = p^{B} \qquad \text{in } U^\mu .$$

On the other hand, $p^{B_\mu} = p$ in B_μ , hence $p^{B} = p$ in $U^\mu \cap B_\mu$. Since p is strict, this means that

$$U^\mu \cap B_\mu \subset B .$$

But $U^\mu \subset U$, and consequently $U^\mu \cap B_\mu \subset U \cap B = \emptyset$. Altogether we have thus proved that $\complement B_\mu = U^\mu$. Note that $B_\mu \setminus B = U \setminus U^\mu$ is finely open and μ-negligible.

Ad b). From the additional hypothesis $\mu(B) = 0$ follows that $\mu(B_\mu) = 0$, and further that the swept-out measure μ^B does not charge the polar sets. In fact, μ^B is carried by B and does not charge any polar set $E \subset B$ in view of §4.7 because $\mu(E) \leq \mu(B) = 0$. Hence μ^B has a fine support by Getoor's theorem (see §8.9). Without making use of this, we proceed to prove that any finely closed set $S \subset \partial_f U^\mu$ such that S carries μ^B, must be equal to $\partial_f U^\mu$. This will complete the proof of b) because $\mu^B = \mu^{B_\mu}$ is indeed carried by $\partial_f B_\mu = \partial_f U^\mu$ according to §4.8, as μ is carried by $\complement B_\mu \subset b(\complement B_\mu)$.

If a finely closed set S carries μ^B , then so does $b(S)$ because $S \setminus b(S)$ is polar. Hence it suffices to consider a <u>base</u> $S \subset \partial_f U^\mu$ such that S carries μ^B , and to prove that $S = \partial_f U^\mu$.

Proceeding by contradiction, we consider any point $y \in \partial_f U^\mu \setminus S$, and choose a regular finely open set W such that

$$y \in W \subset \tilde{W} \subset \complement S .$$

In terms of a strict potential $p \in \mathcal{P}^c$ such that $\int p\,d\mu < +\infty$ we define

$$f = p - p^{\complement W} = p - \hat{R}_p^{\complement W} ,$$

and put $u = f^B$. Note that $f > 0$ in W, whereas $f = 0$ in $\complement W$. Moreover, $\mu^S = \mu^{B_\mu}$ $(= \mu^B)$ by [27, Lemme 3.1] because $S \subset B_\mu$ and S carries μ^B. Since f is the difference between two hyperharmonic functions ≥ 0 in Ω, and these are integrable with respect to μ, we clearly obtain $\int f^A d\mu = \int f\,d\mu^A$ for any set A. (Alternatively we could use the general result in §4.13.) Hence

$$\int u\,d\mu = \int f^B d\mu = \int f\,d\mu^B = \int f\,d\mu^S = \int f^S d\mu = 0 .$$

In fact, $S \subset \complement W$, and so by [27, Lemma 3.1] (or directly)

$$f^S = p^S - (p^{\complement W})^S = 0 .$$

According to Lemma 9.3, $u = f^B$ is finely harmonic (and ≥ 0) in $U = \complement B$. In every fine component V of U^μ we therefore have $u \equiv 0$ by Theorem 12.6 because $\int u\,d\mu = 0$ and $\mu(V) > 0$. This shows that $u = 0$ in all of U^μ, and hence on $\partial_f U^\mu$ because $u = p^B - (p^{\complement W})^B$ is finely continuous. Since $y \in \partial_f U^\mu \subset \partial_f U$, it follows that $u(y) = 0$. On the other hand,

$$u(y) = f^B(y) = f(y) > 0$$

because $y \in B$, and $y \in W = [f > 0]$. This contradiction completes the proof. ‖

Remark. The above theorem was obtained in [27, Théorème 5] for the particular case of a Green space. It was proved further that, in the greenian case, the maximal base $B_\mu = \complement U^\mu$ is the base of the finely closed set

$$A := [G\mu^B = G\mu]$$

and further that

$$A \setminus I(\mu) \subset B_\mu \subset A .$$

Here $G\mu$ and $G\mu^B = \hat{R}^B_{G\mu}$ are the Green potentials of μ and μ^B, respectively, and $I(\mu)$ denotes the polar set $[G\mu = +\infty]$.

An independent, probabilistic proof of Theorem 12.7 for the case $\mu = \varepsilon_x$ was given by Nguyen-Xuan-Loc and Watanabe [35].

Corollary 1. For any admissible measure μ on Ω and any two bases A, B with the corresponding maximal bases A_μ, B_μ,

$$A \subset B \quad \text{implies} \quad A_\mu \subset B_\mu .$$

In fact, $(\complement A)^\mu \supset (\complement B)^\mu$.

Corresponding to the particular case $\mu = \varepsilon_x$, $x \in U$, in Theorem 12.7, we have

Corollary 2. Let x denote a point of a regular finely open set U, and let U^x be the fine component of x in U. Then $B_x := \complement U^x$ is the largest among all bases E such that $\varepsilon_x^E = \varepsilon_x^B$. Moreover, the fine support of ε_x^B is $\partial_f B_x = \partial_f U^x$.

In this corollary, let us now drop the requirement that U be regular. On the other hand, we shall assume for simplicity that U is finely connected. Then we obtain

Corollary 3. Let U be a fine domain, and let x denote either a point of U, or else an irregular fine boundary point for U (that is, a point of $i(\complement U)$). Then the fine support of $\varepsilon_x^{\complement U}$ exists and equals the regular part of the fine boundary of U :

$$\mathrm{supp}_f\, \varepsilon_x^{\complement U} = b(\partial_f U) = b(\complement U) \cap \partial_f U .$$

In fact, the smallest regular finely open set V containing U , that is

$$V = \complement b(\complement U) = U \cup i(\complement U),$$

is a fine domain containing x in either case (§12.5), and $\partial_f V = b(\partial_f U) = b(\complement U) \cap \partial_f U$. Hence the result since $\varepsilon_x^{\complement U} = \varepsilon_x^{b(\complement U)} = \varepsilon_x^{\complement V}$.

This last corollary contains (and is a refinement of) a result established by Effros and Kazdan [24 , Lemma 3.8], asserting that, when x is an irregular boundary point for a relatively compact domain U (in the initial topology on Ω), then the ordinary closed support of $\varepsilon_x^{\complement U}$ contains all the regular boundary points for U .

12.8. We shall now examine the swept-out μ^B of an admissible measure μ onto a base B without assuming that $\mu(B) = 0$, as we did in the second statement of Theorem 12.7. We therefore introduce the traces

$$\mu_B := 1_B \mu ; \qquad \nu := 1_U \mu$$

of μ on B and $U = \complement B$, respectively. Then $\mu = \mu_B + \nu$, and

$$\mu^B = \mu_B + \nu^B$$

because μ_B is carried by B , and so $(\mu_B)^B = \mu_B$. Similarly, $\mu^E = \mu_B + \nu^E$ for any set $E \supset B$, and consequently

$$B_\mu = B_\nu .$$

Since $\nu(B) = 0$, ν^B has the fine support $\partial_f B_\nu = \partial_f B_\mu$.

Now suppose that μ_B possesses a fine support, $\operatorname{supp}_f \mu_B$, (the smallest finely closed set carrying μ_B). Then so does μ^B, and

$$\operatorname{supp}_f \mu^B = \operatorname{supp}_f \mu_B \cup \partial_f B_\mu .$$

On the other hand, without any assumptions concerning the admissible measure μ and the base B , we have the corresponding relation between the quasi supports (which always exist, see §8.9):

$$\operatorname{supp}_q \mu^B = \operatorname{supp}_q \mu_B \cup \partial_f B_\mu .$$

This is clear since ν^B does not charge the polar sets, and hence its quasi support is at the same time its fine support, which equals $\partial_f B_\nu = \partial_f B_\mu$.

Finally, we propose to establish (likewise in complete generality) the following identity

$$\operatorname{supp}_q \mu^B = (B \cap \operatorname{supp}_q \mu) \cup \partial_f B_\mu .$$

(This is not an obvious consequence of the preceding result

because $\operatorname{supp}_f \mu_B$ may be properly contained in $B \cap \operatorname{supp}_f \mu$, and even in the base thereof.) To establish the non trivial inequality $\supset$, let $x \in B \cap \operatorname{supp}_f \mu$, and suppose that $x \in \complement \operatorname{supp}_f \mu_B$. In order to prove that $x \in \partial_f B_\mu$, consider any regular finely open W such that $x \in W \subset \complement \operatorname{supp}_f \mu_B$. The trace $\mu_{W \cap B}$ of μ on $W \cap B$ equals that of μ_B on W, and hence is carried by a polar set. On the other hand, μ_W is not carried by any polar set because $W \cap \operatorname{supp}_f \mu$ is not void (it contains x). It follows that

$$\mu_{W \cap U} = \mu_W - \mu_{W \cap B}$$

is not carried by any polar set. In particular, $\mu(W \cap U) > 0$. Hence $\mu(V) > 0$ for some fine component V of $W \cap U$. The extension of V to a fine component of U is of course likewise non negligible for μ , and hence contained in U^μ. Consequently, $W \cap U^\mu \supset V \neq \emptyset$, showing that indeed $x \in (U^\mu)^\sim$, and hence $x \in \partial_f U^\mu$ because $x \in B = \complement U \subset \complement U^\mu$.

12.9. Theorem. For any finely hyperharmonic function in a finely open set U the following are equivalent:

i) u is not identically $+\infty$ in any fine component of U.

ii) u is finite in some finely dense subset of U.

iii) u is finite quasi everywhere in U.

Proof. The implication ii) $\Longrightarrow$ i) follows from the local connectivity of the fine topology. Next, iii) $\Longrightarrow$ ii) is obvious because a polar set has no finely interior points.

As to the remaining implication i) $\Longrightarrow$ iii), consider first the following case: U is a fine domain, u is defined and finely l.s.c. in $\tilde{U}$ and finely hyperharmonic in U, and moreover $u \geq -p$ in U for some $p \in \mathcal{P}^c$.

Since u is finely continuous in U (Theorem 9.10), the set

$$E := \{ x \in U \mid u(x) = +\infty \}$$

is finely closed relatively to U. To prove that E is polar (under our preliminary assumptions concerning U and u), denote by V any fine component of the finely open set $U \setminus E$ (which is non void by i)), and introduce the fine boundary e of V relative to the finely open set U:

$$e := (\partial_f V) \cap U.$$

Clearly, e is contained in the fine boundary of $U \setminus E$ (or of E) relative to U, that is,

$$e \subset \partial_f (U \setminus E) \cap U \subset E.$$

Fix $x \in V$, and note that, by Theorem 9.4,

$$\int u \, d\varepsilon_x^{\complement V} \leq u(x) < +\infty$$

because $x \in U \setminus E$. It follows that e is negligible with respect to $\varepsilon_x^{\complement V}$ since $u = +\infty$ on e $(\subset E)$. Hence $\varepsilon_x^{\complement V}$ is carried by the finely closed set

$$(\partial_f V) \setminus e = (\partial_f V) \setminus U.$$

According to Corollary 3, §12.7, this shows that $(\partial_f V)\setminus e$ contains $b(\partial_f V)$, and so e must be polar. Now,

$$\emptyset \neq V \subset U\setminus E \subset U\setminus e\,,$$

and $(\partial_f V)\cap(U\setminus e) = \emptyset$. Since e is polar, $U\setminus e$ is a fine domain by Theorem 12.2, and we conclude that $V = U\setminus e$, showing that $E=e$ is indeed polar in the above particular case.

In the general case define E as above, and denote by U' the set of all points of U which are <u>not finely interior to</u> E. For any $x\in U'$ let W_x denote a fine domain such that

$$x \in W_x \subset \widetilde{W}_x \subset U$$

and such that furthermore $u \geqslant -p_x$ in W_x for some $p_x \in \mathcal{P}^c$. (For instance, take for W_x the fine component of x in some set V_x containing x and of class $\mathcal{W}(u^-)$, see §8.4.)

From what was established above follows that u is finite q.e. in W_x, that is, $W_x \cap E$ is polar, the alternative $u \equiv +\infty$ in W_x (that is, $W_x \subset E$) being ruled out since $x\in U'$. In particular, the finely open set W_x cannot meet the fine interior of E, and so $W_x \subset U'$.

This shows that U' is finely open, viz. the union of the finely open sets W_x, $x\in U'$. However, $U\setminus U'$ is likewise finely open, being the fine interior of E. Thus $U\setminus U'$ is either void or else the union of certain fine components of U. By the hypothesis i), no fine component of U is contained in the fine interior $U\setminus U'$ of E, and we conclude that $U'=U$.

We have therefore represented U itself as the union of the finely open sets W_x, $x \in U$. In view of the quasi Lindelöf principle there is a sequence of points $x_n \in U$ and a polar set e' such that

$$U = e' \cup \bigcup_n W_{x_n} ,$$

and consequently E is polar, being the union of the polar set $E \cap e'$ and the sequence of polar sets $E \cap W_{x_n}$. $\square$

The finely hyperharmonic functions with the equivalent properties i), ii), and iii) of the above theorem have been studied in §10 and §11 under the name of finely superharmonic functions.

13. On the balayage of semibounded potentials

For any superharmonic function $p \geq 0$ on Ω we denote by $I(p)$ the polar set $[p = +\infty]$. Also recall that $S_q(p)$ denotes the quasi harmonic support of p (§8.8), and that this equals the fine harmonic support of p if p is finite valued (§9.16).

The notions and results of §13.1 - §13.4 below all carry over mutatis mutandis to the more general case of balayage relative to a given finely open set $U \subset \Omega$ (§11.3), when the semibounded potential p on Ω is replaced by a stable, fine potential p relative to U (§10.9 - §10.12). Naturally, the base of a set $A \subset U$ should be replaced then by the relative base $b(A) \cap U$, and similarly the quasi harmonic support $S_q(p)$ should be taken relatively to U (as in §11.17). Also, p^A should then be read $\hat{R}_p^A$.

13.1. Definition. For any hyperharmonic function $p \geq 0$ on Ω and any base B we write

$$B_p = b([p^B = p]).$$

Recall that $p^B = \hat{R}_p^B$ (§4.12). Clearly $B_p \supset B$. Also note that $p^B = p^{B_p}$, and that B_p is the largest among all bases E such that $p^E = p^B$. In fact, $p^B = p$ on B shows that $B_p \supset B$, hence $p^{B_p} \geq p^B$. On the other hand we have $p^B \geq p$ in $[p^B = p]$, whence

$$p^B \geq p^{[p^B = p]} = p^{B_p}.$$

For any base E such that $p^E = p^B$ we obtain $E_p = B_p$, and so $E \subset B_p$.

From the definition above follows the implication

$$A \subset B \quad \Longrightarrow \quad A_p \subset B_p$$

for any hyperharmonic function $p \geq 0$ and any two bases A, B.

<u>Remark</u>. If Ω is a <u>Green space</u> and $p = G\mu$ the greenian potential of an admissible measure μ, then $B_p = B_{G\mu}$ coincides with the maximal base B_μ studied in Theorem 12.7 above.

13.2. <u>Lemma</u>. Let p denote a superharmonic function ≥ 0 on Ω. With B_p as above, and $I(p) = [p = +\infty]$, we have

$$[p^B = p] \setminus I(p) \subset B_p \subset [p^B = p].$$

<u>Proof</u>. According to Cor. 9.7 (or Theorem 9.13), p^B is finely harmonic off $B \cup I(p)$, in particular off $B_p \cup I(p)$, and hence $p - p^B$ is finely hyperharmonic there by Cor. 9.5. Since $p - p^B \geq 0$, it follows from Theorem 12.6 that the polar set

$$e := [p^B = p] \cap \complement(B_p \cup I(p)) \subset i([p^B = p])$$

is finely open, hence void, being a union of fine components of the finely open set $\complement(B_p \cup I(p))$. - The second inclusion is obvious. ▯

<u>Corollary</u>. For any <u>finite</u>, superharmonic function $p \geq 0$ on Ω and any base B, $[p^B = p]$ is a base ($= B_p$).

13.3. <u>Theorem</u>. Let p denote a <u>semibounded</u> <u>potential</u> on Ω, and B a base. Then $B_p \setminus B$ is the union of those fine components V of $\complement B$ for which p is finely harmonic in $V \setminus I(p)$ (or equivalently: finely harmonic q.e. in V).

Proof. We begin by proving that, for any superharmonic function $p \geq 0$ on Ω, $B_p \setminus B$ is a union of fine components of $\complement B$ (and so is therefore $\complement B_p$). Let V denote any fine component of $\complement B$. By Theorem 12.2, $V \setminus I(p)$ is finely connected because $I(p)$ is polar. Like in the preceding proof, $p - p^B$ is finely hyperharmonic and ≥ 0 in $V \setminus I(p)$, and hence there are only the following two possibilities:

1°) $p = p^B$ in $V \setminus I(p)$, that is, $V \setminus I(p) \subset [p^B = p]$, and consequently $V \subset B_p$ by taking bases:

$$V \subset b(V) = b(V \setminus I(p)) \subset b([p^B = p]) = B_p.$$

2°) $p > p^B$ in $V \setminus I(p)$, that is, $(\complement V) \cup I(p) \supset [p^B = p]$, and consequently $V \subset \complement B_p$:

$$\complement V \supset b(\complement V) = b((\complement V) \cup I(p)) \supset b([p^B = p]) = B_p.$$

Thus any fine component V of $\complement B$ is entirely contained either in $B_p \setminus B$ or in $\complement B_p$, in accordance with the two cases. Note that, in case 1°, p is finely harmonic in $V \setminus I(p)$ $(\subset B_p \setminus B)$ since $p = p^B$ there.

It remains to show that, when p is a semibounded potential, then conversely any fine component V of $\complement B$ such that p is finely harmonic in $V \setminus I(p)$, is contained in $B_p \setminus B$. Now $p^B - p$ is finely hyperharmonic (even finely harmonic) in $V \setminus I(p)$, and finely continuous and $\geq -p$ in $\complement I(p)$. Moreover, $p^B - p = 0$ on $B \setminus I(p)$, in particular q.e. on $\partial_f(V \setminus I(p)) \subset (\partial_f V) \cup I(p)$ because $\partial_f V \subset B$, V being a fine component of $\complement B$. It therefore follows from the fine boundary minimum principle (Theorem 9.1) that $p^B - p \geqslant 0$ in $V \setminus I(p)$, and consequently $p^B = p$ in $V \setminus I(p)$, and indeed in all of V by fine continuity. Thus V is in case 1° above, that is, $V \subset B_p \setminus B$. ▯

Corollary. For any superharmonic function $p \geqslant 0$ on Ω and any base B, $B_p \setminus B$ is finely open, and $\partial_f B_p \subset \partial_f B$.

In fact, $B_p \setminus B$ is finely open, being a union of fine components of the finely open set $\complement B$. Hence $B_p \setminus B$ is contained in the fine interior of B_p, and so is of course the fine interior of B because $B \subset B_p$.

Remark. In order that $B_p = \Omega$, or equivalently $p^B = p$, it is necessary and sufficient that the semibounded potential p be finely harmonic q.e. in $\complement B$, in other words that $S_q(p) \subset B$. This follows from the above theorem, or just as well directly from the fine domination property of p (Theorem 9.2).

The opposite extreme, $B_p = B$, means that p is not finely harmonic q.e. in any (full) fine component of $\complement B$. In particular, $B_p = B$ for all bases B, if and only if $S_q(p) = \Omega$ (cf. in this connection §8.8).

13.4. The following theorem is analogous (and dual) to the general formula for the quasi support of a swept-out measure μ^B expressed in terms of that of the measure μ itself, as obtained at the end of §12.8. In the case of a Green space, this analogy becomes an actual identity in view of the relation $S_q(G\mu) = \operatorname{supp}_q \mu$ for any admissible measure μ (Theorem 8.10).

Theorem. For any semibounded potential p on Ω and any base B, the quasi harmonic support of $p^B = \hat{R}_p^B$ is determined from that of p as follows:

$$S_q(p^B) = (B \cap S_q(p)) \cup \partial_f B_p$$

$$= B \cap [S_q(p) \cup \partial_f B_p].$$

Proof. The two stated expressions for $S_q(p^B)$ are equal because $\partial_f B_p \subset \partial_f B \subset B$ by Cor. 13.3.

As to the inclusion $\subset$, p^B is finely harmonic q.e. off B by Cor. 9.7, and so $S_q(p^B) \subset B$. To establish that $S_q(p^B) \subset S_q(p) \cup \partial_f B_p$, we shall prove that no fine component of the finely open set $\complement[S_q(p) \cup \partial_f B_p]$ can meet $S_q(p^B)$, in other words that p^B is finely harmonic q.e. in any such component V. Now $V \cap S_q(p) = \emptyset$, and so p is finely harmonic q.e. in V. On the other hand, from $V \cap \partial_f B_p = \emptyset$ follows that either $V \subset B_p$ or else $V \subset \complement B_p$. In the former case, p^B coincides in V with p, which is finely harmonic q.e. in V. In the latter case, $V \subset \complement B$, and so p^B is finely harmonic q.e. in V as noted at the beginning of the proof.

To establish the opposite inclusion $\supset$, we shall prove that no fine component V of $\complement S_q(p^B)$ can meet $(B \cap S_q(p)) \cup \partial_f B_p$. Now p^B is finely harmonic q.e. in V, hence everywhere in $V \setminus I(p)$ by Cor. 9.15 because $p^B \leq p < +\infty$ in $V \setminus I(p)$. This shows that $p - p^B$ is finely hyperharmonic (and ≥ 0) in $V \setminus I(p)$. It follows by Theorem 12.6 that either $p = p^B$ in $V \setminus I(p)$, and hence in V, or else $p > p^B$ in $V \setminus I(p)$.

In the former case, $V \subset [p^B = p]$, whence, by taking bases, $V \subset b(V) \subset B_p$. Since V is finely open it follows that $V \cap \partial_f B_p = \emptyset$. Moreover, p is itself finely harmonic q.e. in V , being equal to p^B there, and so

$$V \cap (B \cap S_q(p)) \subset V \cap S_q(p) = \emptyset.$$

In the latter case, $[p^B = p] \subset (\complement V) \cup I(p)$, whence, by taking bases, $B_p \subset b(\complement V) \subset \complement V$, that is $V \cap B_p = \emptyset$, and hence V again does not meet $(B \cap S_q(p)) \cup \partial_f B_p$ $(\subset B_p)$. ▯

13.5. We proceed to interrelate, for any base B , the two dual notions of maximal base - the maximal base $B_p = b([p^B = p])$ defined above for any semibounded potential p , and the maximal base B_μ defined in §12.7 for any admissible measure μ .

<u>Theorem</u>. Let p denote a semibounded potential on Ω , μ an admissible measure on Ω , and B a base. Then μ is carried by B_p if and only if p is finely harmonic in $\complement(B_\mu \cup I(p))$, in other words if and only if $S_q(p) \subset B_\mu$.

<u>Proof</u>. Let (V_n) denote the countable family of those fine components of $\complement B$ which are non negligible with respect to μ . According to Theorem 12.7,

$$\bigcup_n V_n = \complement B_\mu .$$

By Theorem 13.3, each V_n is a fine component either of $B_p \setminus B$ or of $\complement B_p$. Hence μ is carried by B_p if and only if each V_n is one of the fine components of $B_p \setminus B$, that is, if $V_n \subset B_p$. According to the same theorem, this amounts to saying that p should be finely harmonic q.e. in each V_n , or equivalently: q.e. in their union $\complement B_\mu$. ▯

Corresponding to the particular case $\mu = \varepsilon_x$, where we write B_x in place of B_{ε_x} ,(cf. Cor. 2 in §12.7), we obtain

Corollary. For any semibounded potential p on Ω, any $x \in \Omega$, and any base B ,

$$x \in B_p \iff S_q(p) \subset B_x .$$

13.6. As an application of the preceding results we shall establish the following theorem concerning balayage of measures, suggested in the case $\mu = \varepsilon_x$ by Nguyen-Xuan-Loc [35], who also gave an independent, probabilistic proof for that case. - Note that, in its formulation, the theorem has nothing to do with fine harmonicity.

Theorem. For any admissible measure μ and any two bases A and B

$$(\mu^B)^A = \mu^A \iff \mu^{A \cup B} = \mu^B .$$

Proof. We may assume that A and B are bases (otherwise replace them by their bases). The implication $\Longleftarrow$ is obvious on account of [27, Lemme 3.1]:

$$(\mu^B)^A = (\mu^{A \cup B})^A = \mu^A .$$

To establish the opposite implication $\Longrightarrow$, choose a strict potential $q \in \mathcal{P}^c$ in the sense of Constantinescu [21] so that $\int q \, d\mu < +\infty$ (see §3.4 above), and put $p := q^A$. From $(\mu^B)^A = \mu^A$ follows then

$$\int p^B d\mu = \int p \, d\mu ,$$

and so μ is carried by $[p^B = p] = B_p$ (see Cor. 13.2).

From $p \leq q \in \mathcal{P}^{a}$ follows that p is locally bounded, in particular semibounded. Since μ is carried by B_p, we infer from the preceding theorem that p is finely harmonic in $\complement B_\mu$. Hence $q - p$ is finely hyperharmonic (and ≥ 0) in $\complement B_\mu$. Moreover, $[q = p] = A$ because q is strict. It follows now by application of Theorem 12.6 that $A \cap \complement B_\mu$ is a union of fine components of the finely open set $\complement B_\mu$, hence $A \cap \complement B_\mu$ is finely open.

Now q coincides in $A \cap \complement B_\mu$ with p, which is finely harmonic there. Being a strict potential, q is, however, not finely harmonic in any non void, finely open set (§8.8), and consequently $A \cap \complement B_\mu = \emptyset$. Having thus established that $A \subset B_\mu$, and hence $B \subset A \cup B \subset B_\mu$, we conclude that $\mu^{A \cup B} = \mu^{B}$ by the very definition of the maximal base B_μ. ▯

Remark. The analogous result concerning balayage of hyperharmonic functions $p \geq 0$ on Ω:

$$(p^B)^A = p^A \iff p^{A \cup B} = p^B,$$

is more elementary. In fact, when A and B are bases, we obtain from Def. 13.1 and its immediate consequences

$$\begin{aligned} (p^B)^A = p^A &\iff p^B = p \text{ on } A \\ &\iff A \subset B_p \iff A \cup B \subset B_p \\ &\iff p^{A \cup B} = p^B . \end{aligned}$$

In particular, in the greenian case, this leads to an elementary proof of Theorem 13.6 when applied to a Green potential $p = G\mu$.

14. The fine Dirichlet problem

In this section U denotes a finely open subset of the strong harmonic space Ω satisfying axiom (D). The case of an irregular set U will be reduced to the regular case by passage to the smallest regular finely open set V containing U, obtained by adding to U the polar set $i(\complement U) = i(\partial_f U)$ of all irregular points for U (§12.5). Thus

$$V = U \cup i(\complement U) = \complement b(\complement U),$$
$$\partial_f U = \partial_f V \cup i(\complement U),$$
$$b(\partial_f U) = \partial_f V,$$
$$\tilde{U} = \tilde{V}.$$

Also recall that, by Lemma 12.3,

$$\varepsilon_x^{\complement U} = \varepsilon_x^{\complement V} = \varepsilon_x^{\partial_f U} = \varepsilon_x^{\partial_f V} \qquad \text{for } x \in \tilde{U} = \tilde{V}.$$

This harmonic measure is carried by $\partial_f V = b(\partial_f U)$, the regular part of $\partial_f U$.

14.1. Theorem. (The proper fine Dirichlet problem.) Let f denote a real valued, finely continuous function defined on $b(\partial_f U)$. There exists at most one finely continuous extension u of f from $b(\partial_f U)$ to $\tilde{U}$ such that u is finely harmonic in U and majorized there, in absolute value, by some semibounded potential on Ω. If $|f| \leq p$ on $b(\partial_f U)$ for some finite and semibounded potential p on Ω, then the extension in question exists and is given by $u = f^{\complement U}$, that is,

$$u(x) = \int f \, d\varepsilon_x^{\complement U} = \int f \, d\varepsilon_x^{\partial_f U} \qquad \text{for every } x \in \tilde{U}.$$

Proof. The uniqueness follows from the fine boundary minimum principle (Theorem 9.1) applied to $u_1 - u_2$ and $u_2 - u_1$, when u_1 and u_2 denote two extensions of f of the stated kind. The remaining assertions follow immediately from Lemma 9.3 applied to f, defined on the base $\partial_f V = b(\partial_f U)$. In addition, it is found in this way that $|u| \leq p$ in $\tilde{U}$. ▮

14.2. Definition. A finely continuous, real valued function f, defined on the whole fine boundary $\partial_f U$, is called a fine Dirichlet function for U if f has a finely continuous extension u to $\tilde{U}$ such that u is finely harmonic in U and majorized there (hence also in $\tilde{U}$) in absolute value by some finite and semibounded potential.

According to the above theorem, such an extension is always unique. And for the existence of the extension (in other words, for f to be a fine Dirichlet function for U), it is necessary and sufficient that $|f| \leq p$ on $\partial_f U$ for some finite and semibounded potential p on Ω, and further that

$$\int f \, d\varepsilon_x^{\complement U} = f(x) \quad \text{for every} \quad x \in i(\complement U),$$

i.e., that $f^{\complement U} = f$ on the irregular part of the fine boundary of U.

This latter condition, $f^{\complement U} = f$ in $i(\complement U)$ is of course redundant if U is regular.

Note also that the extension in question, if any, must be finely harmonic in the larger finely open set $V = U \cup i(\complement U)$ by virtue of Cor. 9.15, or because $f^{\complement U} = f^{\complement V}$.

<u>Theorem</u>. The class $\mathcal{D}(U)$ of all fine Dirichlet functions for an <u>irregular fine domain</u> U is an <u>antilattice</u> in the sense that $\mathcal{D}(U)$ does not contain the pointwise infimum (or supremum) of two functions $f, g \in \mathcal{D}(U)$ except in the trivial case where $f \leq g$ or $g \leq f$.

<u>Proof</u>. Let $h := \min\{f, g\}$, where $f, g \in \mathcal{D}(U)$, and consider any fixed irregular point x for U. We may assume for instance that $f(x) \leq g(x)$, and hence $h(x) = f(x)$. Then $u := f^{\complement U}$ and $w := h^{\complement U}$ are the extensions of f and h respectively, as described above. It follows that

$$u(x) = f(x) = h(x) = w(x).$$

From $f \geq h$ follows $u \geq w$, and so $u - w$ is finely harmonic and ≥ 0 in $V = U \cup i(\complement U)$. Since U is a fine domain, so is V by Theorem 12.2, and we conclude from Theorem 12.6 that $u - w = 0$, the alternative $u - w > 0$ being ruled out because $u(x) = w(x)$. In particular, $f = h$, that is, $f \leq g$ (on $\partial_f U$). ▯

<u>Remark</u>. This theorem is an extension and refinement of a similar result established by Effros and Kazdan [24, Theorem 3.9] concerning the ordinary Dirichlet problem for a relatively compact, open set (in the initial topology), possessing at least one irregular boundary point.

Observe at this point that every finite and continuous function (in the initial topology), defined on a compact set, is bounded, and hence majorized in absolute value by a semibounded (even by a finite and continuous) potential. For every finite and continuous function f, defined on the usual boundary ∂U of a relatively compact, open set U, the restriction of f to $\partial_f U$ is therefore a fine Dirichlet function for U if f is a usual Dirichlet function for U. This leads to the quoted result from [24].

14.3. In the sequel, f denotes an arbitrary numerical function, defined at least on the fine boundary $\partial_f U$ of the given finely open set U. We shall extend the Perron-Wiener-Brelot method so as to cover the generalized fine Dirichlet problem.

Definition. By a fine superfunction for f (relative to U) we understand a numerical function u, defined and finely hyperharmonic at least in U, such that $u \geq -p$ in U for some finite and semibounded [25] potential p on Ω (which may depend on u), and satisfying the boundary inequality

$$\text{fine}\liminf_{x \to y,\, x \in U} u(x) \geq f(y) \qquad \text{for every } y \in \partial_f U.$$

The pointwise infimum (within $[-\infty, +\infty]^U$) of the family of all fine superfunctions for f relative to U is denoted by

$$\bar{H}_f = \bar{H}_f^U.$$

(Note at this point that the constant $+\infty$ is a fine superfunction for f.)

---

25) The subsequent developments would remain in force if the potential p were required to be finite and continuous, $p \in \mathcal{P}^c$. This will appear from the end of the proof of Theorem 14.6 below. In the particular case where U is relatively compact (in the initial topology), this means that u should be bounded from below in U, just as in the case of the ordinary Dirichlet problem.

The notion of a <u>fine</u> <u>subfunction</u>, v , for f is defined analogously, or equivalently by the requirement that $-v$ should be a fine superfunction for $-f$. Next, $\underline{H}_f = \underline{H}_f^U$ is defined as the pointwise supremum of the family of all fine subfunctions for f , or equivalently

$$\underline{H}_f = -\overline{H}_{-f} .$$

Finally, we may write H_f^U , or simply H_f , in place of $\underline{H}_f$ and $\overline{H}_f$ whenever these are identical.

14.4. <u>Definition</u>. The function f on $\partial_f U$ is called <u>finely</u> <u>resolutive</u> (relatively to U) if

$$-\infty < \underline{H}_f(x) = \overline{H}_f(x) < +\infty \qquad \text{for every } x \in U.$$

14.5. Let again $V = U \cup i(\complement U)$ denote the smallest regular finely open set containing U .

<u>Lemma</u>. For any numerical function f on $\partial_f U$ $(\supset \partial_f V)$

$$\overline{H}_f^U(x) = \overline{H}_f^V(x) ; \qquad \underline{H}_f^U(x) = \underline{H}_f^V(x) \qquad (x \in U).$$

<u>Proof</u>. Let $x_0 \in U$ be given. Since $i(\complement U)$ is polar, there exists an ordinary hyperharmonic function $w \geqslant 0$ on Ω such that $w = +\infty$ everywhere in $i(\complement U)$, whereas $w(x_0) < +\infty$. For any fine superfunction u for f relative to U , and for any number $\varepsilon > 0$, $u + \varepsilon w$ becomes, when given the value $+\infty$ at every point of $i(\complement U)$, a fine superfunction for f relative to V . This follows from Theorem 9.14 because $u + \varepsilon w$ is finely hyperharmonic in $U = V \setminus i(\complement U)$, and finely continuous and $> -\infty$ in all of V . For there is a finite,

semibounded potential p on Ω such that $u \geqslant -p$ in U. Hence $u + \varepsilon w \geqslant -p + \varepsilon w$ $(> -\infty)$ in V, with $-p + \varepsilon w$ finely continuous in Ω and $= +\infty$ in $i(\complement U)$. Consequently, $\bar{H}_f^V \leqslant \bar{H}_f^U + \varepsilon w$ in U, and so $\bar{H}_f^V(x_0) \leqslant \bar{H}_f^U(x_0)$ because $w(x_0) < +\infty$.

On the other hand, for any fine superfunction u for f relative to V, and any finite and semibounded potential p on Ω with $u \geqslant -p$ in V, the function $u + \varepsilon w$, with w and ε as above, is a fine superfunction for f relative to U because again $u + \varepsilon w \geqslant -p + \varepsilon w$ $(\geqslant -p)$ in V, and

$$\operatorname*{fine\,lim}_{z \to y,\, z \in U} (-p(z) + \varepsilon w(z)) = -p(y) + \varepsilon w(y) = +\infty$$

at any point $y \in i(\complement U) = \partial_f U \setminus \partial_f V$. This leads to the opposite inequality $\bar{H}_f^U(x_0) \leqslant \bar{H}_f^V(x_0)$ in the same way as above.

14.6. <u>Theorem</u>. (The generalized fine Dirichlet problem.) Let U denote a finely open set, and f a numerical function defined on $\partial_f U$. Then

$$\bar{H}_f^U(x) = \int^* f \, d\varepsilon_x^{\complement U} \; ; \quad \underline{H}_f^U(x) = \int_* f \, d\varepsilon_x^{\complement U} \quad (x \in U).$$

Each of the functions $\bar{H}_f^U$ and $\underline{H}_f^U$ is therefore finely harmonic in the finely open set of those points of U at which it is finite. In particular, f is finely resolutive if and only if f is integrable with respect to $\varepsilon_x^{\complement U} = \varepsilon_x^{\partial_f U}$ for every $x \in U$. In the affirmative case the function H_f^U is finely harmonic in U, and

$$H_f^U(x) = \int f \, d\varepsilon_x^{\complement U} \qquad \text{for every } x \in U.$$

Proof. According to Lemma 14.5 it suffices to consider the regular case where $\complement U$ and hence $\partial_f U$ are bases. For any fine superfunction u for f relative to U the extension of u to $\tilde{U}$ defined by putting

$$u(y) = \underset{z \to y,\ z \in U}{\text{fine lim inf}}\ u(z) \qquad \text{for every } y \in \partial_f U,$$

is finely l.s.c. in $\tilde{U}$, and $\geq f$ on $\partial_f U$. If p denotes a finite, semibounded potential on Ω such that $u \geq -p$ in U, then the extension of u is $\geq -p$ in all of $\tilde{U}$. For every $x \in U$ it therefore follows from Theorem 9.4 that

$$u(x) \geq \int u\, d\varepsilon_x^{\complement U} \geq \int^* f\, d\varepsilon_x^{\complement U}.$$

This implies that $\bar{H}_f^U(x) \geq \int^* f\, d\varepsilon_x^{\complement U}$.

In proving the opposite inequality for given $x \in U$ (now with $\complement U$ a base), we may suppose that $\int^* f\, d\varepsilon_x^{\complement U} < +\infty$. By definition of the upper integral of numerical functions (see Bourbaki [7, Chap. IV, §4, exerc. 5]) there exists, for any $\varepsilon > 0$, an ordinary l.s.c. function g on Ω, integrable with respect to $\varepsilon_x^{\complement U}$, and such that $g \geq f$ on $\partial_f U$, and $\int g\, d\varepsilon_x^{\complement U} < \int^* f\, d\varepsilon_x^{\complement U} + \varepsilon$.

Let $p > 0$ denote a fixed, finite and continuous potential on Ω, and define a sequence (g_n) by $g_n = \max\{g, -np\}$. Each g_n is then l.s.c. and integrable with respect to $\varepsilon_x^{\complement U}$. Moreover,

$$-np \leq g_n \leq g^+,$$

and $g_n \geq f$ on $\partial_f U$. Finally, $g_n^{\complement U}$ is finite at x, and $g_n^{\complement U} = g_n$ on the base $\complement U$. According to Lemma 9.12, $g_n^{\complement U}$ is therefore a fine superfunction for g_n and hence for f. Since g_n

decreases pointwise to the limit g , we conclude that

$$\bar{H}_f^U(x) \leq \inf_n g_n^{\complement U}(x) = g^{\complement U}(x) < \int^* f \, d\varepsilon_x^{\complement U} + \varepsilon .$$

The remaining assertions of the theorem are now obvious on account of Theorem 9.13. ▯

Remark. For a relatively compact, open set U in the initial topology it follows from the above theorem that $\bar{H}_f^U$ and $\underline{H}_f^U$ coincide with the classical ones, corresponding to the ordinary generalized Dirichlet problem. This was established first by Brelot [9, Théorème 2]. In particular, the fine resolutivity reduces to the ordinary resolutivity in this case.

14.7. In view of the preceding theorem, $\bar{H}_f^U$ and $\underline{H}_f^U$ merely depend on the restriction of f to the regular part $b(\partial_f U)$ of the fine boundary of U. Hence we may allow f to be defined just on $b(\partial_f U)$.

Theorem. Let again U denote a finely open set, and f a numerical function defined on $b(\partial_f U)$. If $|f| \leq p$ on $b(\partial_f U)$ for some finite and semibounded potential p on Ω then

$$\underset{x \to y,\ x \in U}{\text{fine lim}} \bar{H}_f^U(x) = \underset{x \to y,\ x \in U}{\text{fine lim}} \underline{H}_f^U(x) = f(y)$$

holds for any regular fine boundary point $y \in b(\partial_f U)$ at which f is finely continuous (relatively to $b(\partial_f U)$).

Proof. On account of the preceding theorem this follows at once from the remark to Theorem 9.13 applied to $B = b(\partial_f U)$. ▯

More generally, if just $f \geq -p$ on $b(\partial_f U)$, we get in the same way for every $y \in b(\partial_f U)$

$$\underset{x \to y,\ x \in U}{\text{fine lim inf}} \underline{H}_f^U(x) \geq \underset{x \to y,\ x \in b(\partial_f U)}{\text{fine lim inf}} f(x) .$$

15. An application to the study of the Choquet property

Given a numerical function $f \geqslant 0$ on Ω (a "density"), we shall consider the set functions (capacities) $\hat{R}_f^{\cdot}$, $\int \hat{R}_f^{\cdot}\, dm$, and $R_f^{\cdot}(x_0)$, assigning to arbitrary subsets A of Ω the values

$$\hat{R}_f^A \in \mathcal{U}, \qquad \int \hat{R}_f^A\, dm, \qquad \text{and} \qquad R_f^A(x_0),$$

respectively. Here m denotes any given (positive Radon) measure on Ω, and x_0 a given point of Ω.

As usual, $\mathcal{U}$ denotes the complete lattice cone of all hyperharmonic functions $\geqslant 0$ on Ω, endowed with the pointwise order, and $\bigwedge$ designates the infimum in this lattice.

We refer to §5.4 for the definition of a capacity of Choquet's type (that is, possessing the Choquet property).

15.1. <u>Theorem</u>. If the set function $\hat{R}_f^{\cdot}$ (with values in $\mathcal{U}$) is of Choquet's type in the sense that

$$\bigwedge \{ \hat{R}_f^{A \setminus F} \mid F \text{ closed, } F \subset A \} = 0$$

for every finely closed set $A \subset \Omega$, then so is each of the set functions a) $\int \hat{R}_f^{\cdot}\, dm$ and b) $R_f^{\cdot}(x_0)$, provided that it is finite on compact sets.

<u>Proof</u>. Replacing, if necessary, f by $f \cdot 1_K$ for some compact set K, we may assume, in view of the local character of the Choquet property (cf. end of §5.4), that the capacity to be examined is also globally <u>finite</u>, that is,

$$\text{a)} \int \hat{R}_f\, dm < +\infty, \quad \text{resp. b)} \quad R_f(x_0) < +\infty .$$

In case a) we shall write $s := \hat{R}_f$. In case b) we denote by s an arbitrary fixed function of class $\mathcal{U}$ such that $s \geq f$ and $s(x_0) < +\infty$.

Now let A denote a given finely closed set. By hypothesis there is a sequence (cf. §1.1) of closed sets $F_n \subset A$ such that

$$\bigwedge_n u_n = \widehat{\inf_n u_n} = 0 ,$$

when we put $u_n := \hat{R}_f^{A \setminus F_n}$ $(\in \mathcal{U})$. Thus (u_n) decreases pointwise to the limit 0 quasi everywhere. Clearly $u_n \leq \hat{R}_f$.

We begin by considering 3 particular cases.

Case 1°. a) The capacity $\int \hat{R}_f^{\bullet} \, dm$ for m not charging any polar set, resp. b) the capacity $R_f^{\bullet}(x_0)$ for x_0 non polar.

As to a), note that $\int u_n \, dm \longrightarrow 0$ because $u_n \longrightarrow 0$ m-almost everywhere, and $\int u_n \, dm \leq \int \hat{R}_f \, dm < +\infty$.

Clearly, b) reduces to a) because x_0 is non polar, and so

$$R_f^{\bullet}(x_0) = \hat{R}_f^{\bullet}(x_0) = \int \hat{R}_f^{\bullet} \, d\varepsilon_{x_0}$$

with $m := \varepsilon_{x_0}$ not charging any polar set.

Case 2°. a) The capacity $\int \hat{R}_f^{\bullet} \, dm$, now with m carried by the finely open set

$$U := (\complement b(A)) \cap [s < +\infty] ,$$

resp. b) the capacity $R_f^{\bullet}(x_0)$ with x_0 in the analogous set U.

In either case $u_n = \hat{R}_f^{A \setminus F_n}$ is finely harmonic in U by Cor. 9.7. It therefore follows from Theorem 9.11 that $\inf_n u_n$ is likewise finely harmonic, in particular finely continuous in U , and hence $\inf_n u_n = 0$ everywhere in U. Consequently, as to a),

$$\int_\Omega u_n \, dm = \int_U u_n \, dm \to 0 ,$$

and, as to b), $u_n(x_0) \to 0$.

Case 3°. a) The capacity $\int \hat{R}_f^{\cdot} \, dm$ with m carried by a polar subset E of $b(A) \cap [s < +\infty]$, resp. b) the capacity $R_f^{\cdot}(x_0)$ with x_0 a polar point of $b(A) \cap [s < +\infty]$.

Ad a). For any compact set $K \subset E$ we have, denoting by m_K the trace of m on K,

$$\int \hat{R}_f^{A \setminus F} \, dm \leq \int \hat{R}_f^{A \setminus F} \, dm_K + \int_{E \setminus K} s \, dm$$

for every closed set $F \subset A$. Since $\int s \, dm < +\infty$, the second member on the right can be chosen as small as we please by suitable choice of the compact set $K \subset E$, and hence it suffices to show that the first member on the right can be made arbitrarily small by suitable choice of the closed set $F \subset A$. We may therefore assume from the beginning that E itself is compact (otherwise replace m by m_K).

Accordingly, m is now supposed to be carried by a compact, polar set $E \subset b(A) \cap [s < +\infty]$. Since Ω is locally compact and has a countable basis, there exists a decreasing sequence of compact sets K_n containing E in their interiors $\mathring{K}_n$, and such that $\bigcap K_n = E$. Now $A \setminus \mathring{K}_n$ is finely closed, and the set E, carrying m, meets neither $A \setminus \mathring{K}_n$ nor $[s = +\infty]$. We are therefore back in case 2°, now with $A \setminus \mathring{K}_n$ in place of A. Consequently, each $A \setminus \mathring{K}_n$ is quasi closed in the narrow sense (§5.1) with respect to the present capacity $\int \hat{R}_f^{\cdot} \, dm$, and so is therefore

$$A \cup K_n = (A \setminus \mathring{K}_n) \cup K_n ,$$

and finally also the countable intersection

$$\bigcap_n (A \cup K_n) = A \cup \bigcap_n K_n = A \cup E = A .$$

Ad b). The same proof works, now with $E := \{x_0\}$.

We are now in a position to settle the general case.

Ad a). According to Lemma 15.6 below, m admits a (unique) decomposition $m = m_1 + m_0$ such that m_1 does not charge any polar set, whereas m_0 is carried by some polar set. Let m_2 and m_3 denote the traces of m_0 on $\complement b(A)$ and $b(A)$, respectively. Since $\int s\, dm < +\infty$, m does not charge $[s = +\infty]$. Thus we arrive altogether at the following decomposition of m:

$$m = m_1 + m_2 + m_3 ,$$

where m_1 does not charge the polar sets, m_2 is carried by a (polar) subset of $(\complement b(A)) \cap [s < +\infty]$, and m_3 by a polar subset of $b(A) \cap [s < +\infty]$. For $i = 1, 2,$ or 3, we thus are in the case number i considered above, now with m replaced by m_i. For any $\varepsilon > 0$ there exist, accordingly, closed sets $F_i \subset A$ with

$$\int \hat{R}_f^{A \setminus F_i}\, dm_i < \varepsilon .$$

It follows that, for the closed set $F := F_1 \cup F_2 \cup F_3$ $(\subset A)$, we have $\int \hat{R}_f^{A \setminus F}\, dm < 3\varepsilon$.

Ad b). Since $s(x_0) < +\infty$, we are in case 1° if x_0 is non polar, in case 2° if $x_0 \in \complement b(A)$, and in case 3° if x_0 is a polar point of $b(A)$. ▯

15.2. Theorem. For any locally bounded function $f \geqslant 0$ on Ω the capacity $\int \hat{R}^{\cdot}_f \, dm$ is locally finite and of Choquet's type for any admissible measure m.

Proof. Let K denote a compact set, and choose $\varphi \in \mathcal{C}_0^+(\Omega)$ so that $f \leqslant \varphi$ on K. Then $\hat{R}_f^K \leqslant \hat{R}_\varphi \in \mathcal{P}_0^c$ (§1.9), and so

$$\int \hat{R}_f^K \, dm \leqslant \int \hat{R}_\varphi \, dm < +\infty$$

because m is admissible (Def. 3.1). In view of its local character (see end of §5.4), the Choquet property for $\hat{R}^{\cdot}_1$ as established by Brelot [11], [14] (see also §5.4 above) implies the same for the capacity $\hat{R}^{\cdot}_f$ for any locally bounded $f \geqslant 0$, and hence also for the capacity $\int \hat{R}^{\cdot}_f \, dm$ on account of the preceding theorem. ∎

15.3. Remark. In the classical case of a Green space, and more generally in case A_2 of Brelot's axiomatic theory (cf. §2.7 above), it was proved by Brelot [14] that the $\mathcal{U}$-valued capacity $\hat{R}^{\cdot}_f$ is of Choquet's type (cf. Theorem 15.1 above) provided just that f be majorized q.e. on every compact set by some superharmonic function on Ω.

Under these circumstances, each of the capacities $\int \hat{R}^{\cdot}_f \, dm$ and $R^{\cdot}_f\,(x_0)$ is therefore likewise of Choquet's type, by Theorem 15.1, provided that it is locally finite (that is, finite on every compact set), e.g. if m is a harmonic measure, resp. x_0 non polar.

15.4. Remark. Even in the greenian case and with f superharmonic (even a semibounded potential) on Ω, the Choquet property of $\int \hat{R}^{\cdot}_f \, dm$ and of $R^{\cdot}_f(x_0)$ may go lost if we drop the hypothesis of local finiteness for these capacities. This appears from the simple example considered in §8.10, viz. $f = G\mu = G\varepsilon_x^A$, where A is a finely closed set, thin at the point x (not in A), and such that, nevertheless, $f(x) = +\infty$. It is possible to choose A so that, furthermore, $A \cup \{x\}$ is compact (cf. note 21, p. 74). For any closed (hence compact) set $F \subset A$ we then have $x \in \complement(A \setminus F)$, and hence by Lemma 4.5

$$R_f^{A \setminus F}(x) = \hat{R}_f^{A \setminus F}(x) = +\infty .$$

In fact, $\mu^A = \mu$, and so

$$\hat{R}_f^A(x) = (G\mu^A)(x) = G\mu(x) = +\infty ,$$

whereas, on the other hand, $R_f^F(x) < +\infty$ because $f = G\varepsilon_x$ on F, and $G\varepsilon_x$ is finite and continuous on the compact set F, hence bounded there. - This shows that $\int \hat{R}_f^{\cdot}\, dm$ (with $m = \varepsilon_x$) and $R_f^{\cdot}(x)$ do <u>not</u> have the Choquet property. The former capacity (but not the latter) has, however a weaker property which we proceed to describe.

15.5. Recall the definition of a quasi closed set (§5.1), and the fact that, in the case of an <u>outer</u> capacity, any quasi closed set is even quasi closed in the narrow sense occurring in the definition of the Choquet property.

<u>Theorem</u>. Let m denote a measure carried by a <u>compact</u>, <u>polar</u> set E, and $f \geq 0$ a function such that $\hat{R}_f^K$ is superharmonic for every compact set K, and such that the capacity $\hat{R}_f^{\cdot}$ is of Choquet's type (e.g. Ω a Green space). Under these hypotheses every finely closed set is quasi closed also with respect to the capacity $\int \hat{R}_f^{\cdot}\, dm$.

Proof. We may again assume that $\hat{R}_f$ is itself superharmonic in Ω. As in the proof of case 3^0 of Theorem 15.1, choose a sequence of compact sets K_n covering Ω so that $\mathring{K}_n \supset E$. Then $A \setminus \mathring{K}_n$ is finely closed, and hence there is a sequence (§1.1) of closed sets $F_j \subset A \setminus \mathring{K}_n$ with the property that

$$\bigwedge_j \hat{R}_f^{(A \setminus \mathring{K}_n) \setminus F_j} = 0 .$$

Since $\hat{R}_f$ is superharmonic in Ω, $u_j := \hat{R}_f^{(A \setminus \mathring{K}_n) \setminus F_j}$ is harmonic in $\mathring{K}_n$. So is therefore $\inf_j u_j$, which thus vanishes everywhere in $\mathring{K}_n$. Since m is carried by the compact set $E \subset \mathring{K}_n$, we find that $\int u_j \, dm < +\infty$, and consequently $\inf_j \int u_j \, dm = 0$. This shows that $A \setminus \mathring{K}_n$ is quasi closed (even in the narrow sense) with respect to $\int \hat{R}_f^{\cdot} \, dm$, and hence so is $A \cup E$ (cf. the proof of case 3^0 of Theorem 15.1). Since E is polar, A is equivalent to $A \cup E$, and consequently A itself is quasi closed (though not necessarily in the narrow sense, as shown by the example in §15.4). ▯

Corresponding to the particular case $m = \varepsilon_{x_0}$ we have, even for a polar point x_0, the following corollary (which, for a non polar point would be contained in Remark 15.3 because a superharmonic function is always finite at any non polar point):

Corollary. Every finely closed set is quasi closed with respect to the capacity $\hat{R}_f^{\cdot}(x_0)$ provided that $\hat{R}_f^K$ is superharmonic for every compact set K and that Ω is, say, a Green space.

15.6. We denote by $\mathcal{M}$ the conditionally complete pre vector lattice of all (positive Radon) measures on Ω . In view of the application in the proof of Theorem 15.1 above we shall establish the following lemma.

Lemma. The following two subsets $\mathcal{M}_r$ and $\mathcal{M}_s$ of $\mathcal{M}$ are complementary orthogonal bands in $\mathcal{M}$:

$$\mathcal{M}_r := \{ \mu \in \mathcal{M} \mid \mu \text{ does not charge any polar set} \},$$

$$\mathcal{M}_s := \{ \mu \in \mathcal{M} \mid \mu \text{ is carried by some polar set} \}.$$

Thus every $\mu \in \mathcal{M}$ admits a unique decomposition of the form $\mu = \mu_r + \mu_s$ with $\mu_r \in \mathcal{M}_r$ and $\mu_s \in \mathcal{M}_s$.

Proof. The proof is based on the fact that every majorized, upper directed subset of $\mathcal{M}$ contains a countable set with the same supremum within $\mathcal{M}$. In view of this it is easily verified that $\mathcal{M}_r$ and $\mathcal{M}_s$ are bands in $\mathcal{M}$, and they are clearly orthogonal to one another, $\mathcal{M}_r \cap \mathcal{M}_s = \{0\}$. Finally, any measure μ such that $\inf(\mu, \nu) = 0$ for all $\nu \in \mathcal{M}_s$, must belong to $\mathcal{M}_r$. In fact, if $\mu^*(E) > 0$ for some polar set E (which we may assume to be μ-measurable since every polar set is contained in a polar G_δ), then the trace ν of μ on E would satisfy $\nu \in \mathcal{M}_s$, $\nu \leq \mu$, and $\nu \neq 0$, which is impossible. ▮

References

1. H. Bauer: Harmonische Räume und ihre Potentialtheorie. Lecture Notes in Mathematics, 22. Berlin 1966.

2. C. Berg: Quelques propriétés de la topologie fine dans la théorie du potentiel et des processus standard. Bull. Sc. Math. (2), 95 (1971), 27 - 31.

3. N. Boboc, C. Constantinescu, and A. Cornea: Axiomatic theory of harmonic functions. Non-negative superharmonic functions. Ann. Inst. Fourier, 15.1 (1965), 283 - 312.

4. N. Boboc, C. Constantinescu, and A. Cornea: Axiomatic theory of harmonic functions. Balayage. Ann. Inst. Fourier, 15.2 (1965), 37 - 70.

5. N. Boboc et A. Cornea: Espaces harmoniques. Axiome D et théorème de convergence. Rev. Roum. Math. Pures et Appl., 13 (1968), 933 - 947.

6. N. Bourbaki: Éléments de Mathématique. Topologie générale. Paris 1958-61.

7. N. Bourbaki: Éléments de Mathématique. Intégration. Paris 1952-59.

8. M. Brelot: Lectures on Potential Theory. Tata Institute of Fundamental Research. Bombay 1960.

9. M. Brelot: Sur l'allure des fonctions harmoniques et surharmoniques à la frontière. Math. Nachr., 4 (1950-51), 298 - 307.

10. M. Brelot: Introduction axiomatique de l'effilement. Ann. Mat. Pura ed Appl. (4), 57 (1962), 77 - 96.

11. M. Brelot: Intégrabilité uniforme. Quelques applications à la théorie du potentiel. Sém. Théorie du Potentiel, 6e année (1962), n° 1a.

12. M. Brelot: Quelques propriétés et applications nouvelles de l'effilement. Sém. Théorie du Potentiel, 6e année (1962), n° 1c.

13. M. Brelot: Axiomatique des fonctions harmoniques. Montréal 1966.

14. M. Brelot: Recherches axiomatiques sur un théorème de Choquet concernant l'effilement. Nagoya Math. J., 30 (1967), 39 - 46.

15. M. Brelot: Capacity and balayage for decreasing sets. Proc. Fifth Berkeley Symp. Math. Stat. and Prob. Berkeley 1966, 279 -293.

16. M. Brelot: La topologie fine en théorie du potentiel. Lecture Notes in Mathematics, 31. Berlin 1967, 36 - 47.

17. M. Brelot: On Topologies and Boundaries in Potential Theory. Lecture Notes in Mathematics, 175. Berlin 1971.

18. H. Cartan: Théorie générale du balayage en potentiel newtonien. Ann. Univ. Grenoble, 22 (1946), 221 - 280.

19. G. Choquet: Sur les fondemants de la théorie fine du potentiel. Sém. Théorie du Potentiel, 1re année (1957), n° 1.

20. G. Choquet: Sur les points d'effilement d'un ensemble. Application à l'étude de la capacité. Ann. Inst. Fourier, 9 (1959), 91 - 102.

21. C. Constantinescu: Some properties of the balayage of measures on a harmonic space. Ann. Inst. Fourier, 17.1 (1967), 273 - 293.

22. C. Constantinescu and A. Cornea: Examples in the theory of harmonic spaces. Seminar über Potentialtheorie. Lecture Notes in Mathematics, 69 (1968), 161 - 171.

22 bis. C. Constantinescu and A. Cornea: Potential Theory on Harmonic Spaces. (To appear.)

23. J.L. Doob: Applications to analysis of a topological definition of smallness of a set. Bull. Amer. Math. Soc., 72 (1966), 579 - 600.

24. E.G. Effros and J.L. Kazdan: Applications of Choquet simplexes to elliptic and parabolic boundary value problems. J. Diff. Eqs., 8 (1970), 95 - 134.

25. B. Fuglede: The quasi topology associated with a countably subadditive set function. Ann. Inst. Fourier, 21.1 (1971), 123 - 169.

26. B. Fuglede: Capacity as a sublinear functional generalizing an integral. Mat. Fys. Medd. Dan. Vid. Selsk., 38 (1971), n° 7.

27. B. Fuglede: Connexion en topologie fine et balayage des mesures. Ann. Inst. Fourier, 21.3 (1971),

28. B. Fuglede: Fine Connectivity and finely harmonic functions. Actes Congr. Internat. Mathem. Nice (1970), II, 513 - 519.

29. R.-M. Hervé: Recherches axiomatiques sur la théorie des fonctions surharmoniques et du potentiel. Ann. Inst. Fourier, 12 (1962), 415 - 571.

30. J. Köhn und M. Sieveking: Reguläre und extremale Randpunkte in der Potentialtheorie. Revue Roum. Math. Pures et Appl., 12 (1967), 1489 - 1502.

31. G. Mokobodzki: Cônes de potentiels et noyaux subordonnés. (Potential theory, C.I.M.E., Stresa 1969, 207 - 248.)

32. Nguyen-Xuan-Loc: Fine boundary minimum principle and dual process. (Preprint, Aarhus 1971-72).

33. Nguyen-Xuan-Loc: Characterization of excessive functions on finely open, nearly Borel sets. Math. Ann. 196 (1972), 250 - 268.

34. Nguyen-Xuan-Loc: Sur les potentiels semi-bornés. C.R. Acad. Sc. Paris, 274 (1972), 767 - 770.

35. Nguyen-Xuan-Loc and T. Watanabe: Characterization of fine domains for a certain class of Markov processes with applications to Brelot harmonic spaces. Z. f. Wahrscheinlichkeitstheorie u. verw. Gebiete, 21 (1972), 167 - 178.

36. J.C. Taylor: Duality and the Martin compactification. (To appear in Annales de l'Institut Fourier.)

Vol. 178: Th. Bröcker und T. tom Dieck, Kobordismentheorie. XVI, 191 Seiten. 1970. DM 18,–

Vol. 179: Seminaire Bourbaki – vol. 1968/69. Exposés 347-363. IV, 295 pages. 1971. DM 22,–

Vol. 180: Séminaire Bourbaki – vol. 1969/70. Exposés 364-381. IV, 310 pages. 1971. DM 22,–

Vol. 181: F. DeMeyer and E. Ingraham, Separable Algebras over Commutative Rings. V, 157 pages. 1971. DM 16.

Vol. 182: L. D. Baumert. Cyclic Difference Sets. VI, 166 pages. 1971. DM 16,–

Vol. 183: Analytic Theory of Differential Equations. Edited by P. F. Hsieh and A. W. J. Stoddart. VI, 225 pages. 1971. DM 20,–

Vol. 184: Symposium on Several Complex Variables, Park City, Utah, 1970. Edited by R. M. Brooks. V, 234 pages. 1971. DM 20,–

Vol. 185: Several Complex Variables II, Maryland 1970. Edited by J. Horváth. III, 287 pages. 1971. DM 24,–

Vol. 186: Recent Trends in Graph Theory. Edited by M. Capobianco/ J. B. Frechen/M. Krolik. VI, 219 pages. 1971. DM 18,–

Vol. 187: H. S. Shapiro, Topics in Approximation Theory. VIII, 275 pages. 1971. DM 22,–

Vol. 188: Symposium on Semantics of Algorithmic Languages. Edited by E. Engeler. VI, 372 pages. 1971. DM 26,–

Vol. 189: A. Weil, Dirichlet Series and Automorphic Forms. V, 164 pages. 1971. DM 16,–

Vol. 190: Martingales. A Report on a Meeting at Oberwolfach, May 17-23, 1970. Edited by H. Dinges. V, 75 pages. 1971. DM 16,–

Vol. 191: Séminaire de Probabilités V. Edited by P. A. Meyer. IV, 372 pages. 1971. DM 26,–

Vol. 192: Proceedings of Liverpool Singularities – Symposium I. Edited by C. T. C. Wall. V, 319 pages. 1971. DM 24,–

Vol. 193: Symposium on the Theory of Numerical Analysis. Edited by J. Ll. Morris. VI, 152 pages. 1971. DM 16,–

Vol. 194: M. Berger, P. Gauduchon et E. Mazet. Le Spectre d'une Variété Riemannienne. VII, 251 pages. 1971. DM 22,–

Vol. 195: Reports of the Midwest Category Seminar V. Edited by J.W. Gray and S. Mac Lane.III, 255 pages. 1971. DM 22,–

Vol. 196: H-spaces – Neuchâtel (Suisse)- Août 1970. Edited by F. Sigrist, V, 156 pages. 1971. DM 16,–

Vol. 197: Manifolds – Amsterdam 1970. Edited by N. H. Kuiper. V, 231 pages. 1971. DM 20,–

Vol. 198: M. Hervé, Analytic and Plurisubharmonic Functions in Finite and Infinite Dimensional Spaces. VI, 90 pages. 1971. DM 16,–

Vol. 199: Ch. J. Mozzochi, On the Pointwise Convergence of Fourier Series. VII, 87 pages. 1971. DM 16,–

Vol. 200: U. Neri, Singular Integrals. VII, 272 pages. 1971. DM 22,–

Vol. 201: J. H. van Lint, Coding Theory. VII, 136 pages. 1971. DM 16,–

Vol. 202: J. Benedetto, Harmonic Analysis on Totally Disconnected Sets. VIII, 261 pages. 1971. DM 22,–

Vol. 203: D. Knutson, Algebraic Spaces. VI, 261 pages. 1971. DM 22,–

Vol. 204: A. Zygmund, Intégrales Singulières. IV, 53 pages. 1971. DM 16,–

Vol. 205: Séminaire Pierre Lelong (Analyse) Année 1970. VI, 243 pages. 1971. DM 20,–

Vol. 206: Symposium on Differential Equations and Dynamical Systems. Edited by D. Chillingworth. XI, 173 pages. 1971. DM 16,–

Vol. 207: L. Bernstein, The Jacobi-Perron Algorithm – Its Theory and Application. IV, 161 pages. 1971. DM 16,–

Vol. 208: A. Grothendieck and J. P. Murre, The Tame Fundamental Group of a Formal Neighbourhood of a Divisor with Normal Crossings on a Scheme. VIII, 133 pages. 1971. DM 16,–

Vol. 209: Proceedings of Liverpool Singularities Symposium II. Edited by C. T. C. Wall. V, 280 pages. 1971. DM 22,–

Vol. 210: M. Eichler, Projective Varieties and Modular Forms. III, 118 pages. 1971. DM 16,–

Vol. 211: Théorie des Matroïdes. Edité par C. P. Bruter. III, 108 pages. 1971. DM 16,–

Vol. 212: B. Scarpellini, Proof Theory and Intuitionistic Systems. VII, 291 pages. 1971. DM 24,–

Vol. 213: H. Hogbe-Nlend, Théorie des Bornologies et Applications. V, 168 pages. 1971. DM 18,–

Vol. 214: M. Smorodinsky, Ergodic Theory, Entropy. V, 64 pages. 1971. DM 16,–

Vol. 215: P. Antonelli, D. Burghelea and P. J. Kahn, The Concordance-Homotopy Groups of Geometric Automorphism Groups. X, 140 pages. 1971. DM 16,–

Vol. 216: H. Maaß, Siegel's Modular Forms and Dirichlet Series. VII, 328 pages. 1971. DM 20,–

Vol. 217: T. J. Jech, Lectures in Set Theory with Particular Emphasis on the Method of Forcing. V, 137 pages. 1971. DM 16,–

Vol. 218: C. P. Schnorr, Zufälligkeit und Wahrscheinlichkeit. IV, 212 Seiten 1971. DM 20,–

Vol. 219: N. L. Alling and N. Greenleaf, Foundations of the Theory of Klein Surfaces. IX, 117 pages. 1971. DM 16,–

Vol. 220: W. A. Coppel, Disconjugacy. V, 148 pages. 1971. DM 16,–

Vol. 221: P. Gabriel und F. Ulmer, Lokal präsentierbare Kategorien. V, 200 Seiten. 1971. DM 18,–

Vol. 222: C. Meghea, Compactification des Espaces Harmoniques. III, 108 pages. 1971. DM 16,–

Vol. 223: U. Felgner, Models of ZF-Set Theory. VI, 173 pages. 1971. DM 16,–

Vol. 224: Revètements Etales et Groupe Fondamental. (SGA 1). Dirigé par A. Grothendieck XXII, 447 pages. 1971. DM 30,–

Vol. 225: Théorie des Intersections et Théorème de Riemann-Roch. (SGA 6). Dirigé par P. Berthelot, A. Grothendieck et L. Illusie. XII, 700 pages. 1971. DM 40,–

Vol. 226: Seminar on Potential Theory, II. Edited by H. Bauer. IV, 170 pages. 1971. DM 18,–

Vol. 227: H. L. Montgomery, Topics in Multiplicative Number Theory. IX, 178 pages. 1971. DM 18,–

Vol. 228: Conference on Applications of Numerical Analysis. Edited by J. Ll. Morris. X, 358 pages. 1971. DM 26,–

Vol. 229: J. Väisälä, Lectures on n-Dimensional Quasiconforma Mappings. XIV, 144 pages. 1971. DM 16,–

Vol. 230: L. Waelbroeck, Topological Vector Spaces and Algebras. VII, 158 pages. 1971. DM 16,–

Vol. 231: H. Reiter, L^1-Algebras and Segal Algebras. XI, 113 pages 1971. DM 16,–

Vol. 232: T. H. Ganelius, Tauberian Remainder Theorems. VI, 7 pages. 1971. DM 16,–

Vol. 233: C. P. Tsokos and W. J. Padgett. Random Integral Equation with Applications to Stochastic Systems. VII, 174 pages. 1971. D 18,–

Vol. 234: A. Andreotti and W. Stoll. Analytic and Algebraic Deper dence of Meromorphic Functions. III, 390 pages. 1971. DM 26,

Vol. 235: Global Differentiable Dynamics. Edited by O. Hájek, A. Lohwater, and R. McCann. X, 140 pages. 1971. DM 16,–

Vol. 236: M. Barr, P. A. Grillet, and D. H. van Osdol. Exact Cat gories and Categories of Sheaves. VII, 239 pages. 1971, DM 20

Vol. 237: B. Stenström. Rings and Modules of Quotients. VII, 1 pages. 1971. DM 16,–

Vol. 238: Der kanonische Modul eines Cohen-Macaulay-Rings. H ausgegeben von Jürgen Herzog und Ernst Kunz. VI, 103 Seit 1971. DM 16,–

Vol. 239: L. Illusie, Complexe Cotangent et Déformations I. XV, 3 pages. 1971. DM 26,–

Vol. 240: A. Kerber, Representations of Permutation Groups I. 192 pages. 1971. DM 18,–

Vol. 241: S. Kaneyuki, Homogeneous Bounded Domains and Sie Domains. V, 89 pages. 1971. DM 16,–

Vol. 242: R. R. Coifman et G. Weiss, Analyse Harmonique N Commutative sur Certains Espaces. V, 160 pages. 1971. DM 1

Vol. 243: Japan-United States Seminar on Ordinary Differential Functional Equations. Edited by M. Urabe. VIII, 332 pages. 19 DM 26,–

Vol. 244: Séminaire Bourbaki – vol. 1970/71. Exposés 382– IV, 356 pages. 1971. DM 26,–

Vol. 245: D. E. Cohen, Groups of Cohomological Dimension Or 99 pages. 1972. DM 16,–